STORM

The Earth series traces the historical significance and cultural history of natural phenomena. Written by experts who are passionate about their subject, titles in the series bring together science, art, literature, mythology, religion and popular culture, exploring and explaining the planet we inhabit in new and exciting ways.

Series editor: Daniel Allen

In the same series

Air Peter Adey
Cave Ralph Crane and Lisa Fletcher
Desert Roslynn D. Haynes
Earthquake Andrew Robinson
Fire Stephen J. Pyne
Flood John Withington
Gold Rebecca Zorach and Michael W. Phillips Jr
Islands Stephen A. Royle
Lightning Derek M. Elsom
Meteorite Maria Golia
Moon Edgar Williams
Mountain Veronica della Dora
South Pole Elizabeth Leane
Storm John Withington
Tsunami Richard Hamblyn
Volcano James Hamilton
Water Veronica Strang
Waterfall Brian J. Hudson

Storm

John Withington

REAKTION BOOKS

Published by Reaktion Books Ltd
Unit 32, Waterside
44–48 Wharf Road
London N1 7UX, UK
www.reaktionbooks.co.uk

First published 2016

Printed and bound in China

A catalogue record for this book is available from the British Library

ISBN 978 1 78023 661 2

CONTENTS

Introduction 7

1 Religion 11

2 Nature 35

3 Effects 69

4 Events 87

5 Literature 107

6 Spectacle 129

7 Futures 159

NOTABLE STORMS 173
REFERENCES 175
SELECT BIBLIOGRAPHY 181
ASSOCIATIONS AND WEBSITES 183
PHOTO ACKNOWLEDGEMENTS 185
INDEX 186

Introduction

Nothing, wrote the English author Jerome K. Jerome in 1908, was 'more beautiful than the love that has weathered the storms of life', while a century earlier, in 'The Bride of Abydos', the poet Lord Byron has his hero entreating his beloved: 'Be thou the rainbow to the storms of life!'

It is no surprise to see storms used in this way as a symbol for the thousand natural shocks of our time on earth. Every one of us will at some time experience this sudden, awesome, sometimes terrifying demonstration of nature's power – as blizzard, hailstorm, thunder and lightning, drenching rain, dust or sandstorm, tornado or hurricane. Christianity is permeated with storm metaphors. The co-founder of Methodism, Charles Wesley, wrote a hymn asking Jesus to shield us 'till the storm of life is past', while for another composer, Sir Robert Grant, writing in the nineteenth century, a storm was the dreadful manifestation of God's power:

> His chariots of wrath
> The deep thunderclouds form,
> And dark is his path
> On the wings of the storm.

The wind was 'all bangs, howls, rattles'. A tropical storm.

As far as the cult pop group The Doors were concerned, all of us were 'Riders on the Storm', thrown into our turbulent lives. This resonant piece of apocalyptic rock was the last

A tornado at Dimmitt, Texas, 2 June 1995.

track ever recorded by the band with their controversial lead singer, Jim Morrison, and was said to have entered the hit parade the day he died, aged 27. The Rolling Stones' 'Jumping Jack Flash' was born in a 'crossfire hurricane', while the fourth movement of Beethoven's 'Pastoral' Symphony is entitled 'Storm and Tempest'. Many musicians have taken names that allude to storms – Lightnin' Hopkins, The Tornados, Johnny and the Hurricanes. Among sporting heroes, we have had snooker player Alex 'Hurricane' Higgins, boxer Rubin 'Hurricane' Carter, and fast bowler Frank 'Typhoon' Tyson. Not surprisingly the language of war has also been colonized – General 'Stormin' Norman' Schwarzkopf led the invasion of Kuwait, which was code-named Desert Storm. Fighter aircraft have been dubbed Tornado, Typhoon and Lightning.

Along with Jerome and Byron, many other writers have drawn on storm imagery, like John Webster in *The White Devil*:

> My soul, like to a ship in a black storm,
> Is driven, I know not whither.

Winston Churchill called the first volume of his history of the Second World War *The Gathering Storm*. And storm metaphors permeate everyday sayings. People can be subjected to a hail of insults. Misfortune strikes like lightning. We stir up a storm of controversy, or win thunderous applause.

This book tells the story of storms – how they turn the lives of some upside-down, and how others need them to survive. How they have changed the course of history, and played crucial roles in religion, art, films and literature. How humans have tried to observe and even control them, and how our relationship with them may change in the future.

I Religion

'When it storms I freak and don't know how to deal with it.' 'The lightning alone makes my skin crawl.' 'I tried to be rational, and realise the thunder was noise and wasn't going to kill me.' These are just a few of the heartfelt howls of anguish on the Internet from people who are terrified of storms. A senior meteorologist confided that he 'was petrified at the sound of thunder or the flash of lightning', and that this was what had motivated him to pursue his career.[1] If we modern humans, with so much scientific knowledge at our disposal, can be so scared of storms, what effect must their sudden, arbitrary, fearsome power have had on our ancestors, who did not have any rational apparatus for understanding them?

It is no wonder that in many early religions the principal deity is the god of thunder. Zeus, for example, chief of the ancient Greek pantheon, was often portrayed brandishing a thunderbolt, and was regarded as the sender of wind, rain, lightning and thunder. His voice was said to be heard in its crash, and places that were frequently struck by lightning were fenced off and consecrated to him. According to some interpretations, the Elysian Fields, the final home of ancient Greek heroes, derived their name from lightning. The ancient Greeks considered the oak the tree most often struck by lightning (just as an old English saying warns: 'Beware of an oak, it draws the stroke'[2]) and a particular specimen at Dodona, where thunderstorms were supposed to rage more frequently than anywhere else in Europe, was said to be revered by Zeus. The chief Greek god's junior

An English oak. The tree was revered in some ancient religions because it is often struck by lightning.

colleague Boreas, the god of the north wind, was an aggressive spirit with a violent temper who was known as the 'Devouring One'.[3] Boreas was said to mate with mares, and a superstition grew that if a mare turned her hindquarters to face the north wind, she could conceive without any help from a stallion. Boreas tried to woo the nymph Oreithyia, but was incapable of breathing gently, so lovers' sighs were out of the question, and he made little progress. Finally he reverted to type and abducted her. Sometimes the Greeks found Boreas's gales useful. When Athens was threatened by the Persians in 480 BC, he blew and wrecked the enemy fleet. At other times his interventions were less welcome. When Aeneas and his followers were escaping from Troy after the city's destruction, Zeus' wife Hera spotted them. She still harboured a grudge against the Trojans because Paris had passed over her in the goddess's beauty contest in favour of Aphrodite. Now she got Boreas and other winds to whip up a terrible storm so that Aeneas' ships were in danger of being wrecked, until Zeus' brother, Poseidon, quietened the tempest and calmed the sea, and the Trojans were able to find refuge at Carthage.

The Greeks' reverence for their thunder god may have been influenced by older religions among people with whom they had close ties, such as the Hittites, who conquered Anatolia and much of Asia Minor in the second millennium BC. One of their chief deities was the storm god Tarhun. Like Zeus his symbol was a thunderbolt, though he also carried a club or an axe. Tarhun appears to have been derived from an even earlier eminent storm god, Teshub, worshipped by the Hurrians, who were first heard of in Mesopotamia (modern-day Iraq) in the third millennium before Christ. Teshub is sometimes depicted as driving a chariot drawn by bulls. Indeed, storm gods could be found all over the Middle East, and the chief Roman god, Jupiter, was also the god of thunder. Jupiter was a Zeus clone; anywhere struck by lightning was made his property.

This worship of storm gods illustrates the ambivalent relationship humanity has always had with tempests. They may bring fear, destruction and even death, but they also confer

Peter Paul Rubens, *Boreas Abducting Oreithyia*, *c.* 1615, oil on wood.

benefits, like rain, that are essential for life. In the Mesopotamian city of Girsu, Ninurta was the god not just of thunder, but of the plough. His name meant 'Rain Cloud' and he was sometimes depicted as a thundercloud in the form of an enormous black bird with a lion's head, floating on outstretched wings. Ninurta's father was Enlil, whose name meant 'Lord Wind'. Fierce hurricanes as well as gentle breezes were supposed to emanate from his mouth. Enlil was the gods' enforcer: he carried out their decrees. According to one ancient text, it was he who was responsible for the original separation of heaven and earth, while in the *Epic of Gilgamesh*, a Mesopotamian story dating back perhaps as far as the third millennium before Christ, he finds himself called upon to execute a terrible task. According to the story, human beings had become so numerous that the noise they made was unbearable, and the gods could no longer sleep. So they decided to 'exterminate mankind'. Ninurta 'turned daylight to darkness' and summoned up a storm and a half: 'the tempest raged, gathering fury as it went, it poured over the people like the tides of battle; a man could not see his brother nor the people be seen from heaven.' Even the gods were terrified. 'For six days and six nights the winds blew, torrent and tempest and flood overwhelmed the world.' Then on the seventh day, the storm subsided, and the sea 'stretched as flat as a roof-top'. Everywhere, 'there was silence, all mankind was turned to clay'. Well not quite. Unbeknownst to Enlil, one of the gods had tipped off a man named Utnapishtim about the impending catastrophe, and he and his family had commissioned a boat and escaped. With them they took samples of 'the beast of the field, both wild and tame, and all the craftsmen' who had built their vessel. At first, Enlil was furious. 'No one was to have survived the destruction!', he raged, but eventually he relented, and even conferred immortality on Utnapishtim and his wife.[4] Enlil continued to be extolled as an important deity until well into the first millennium BC, but he was replaced as the gods' enforcer by Marduk, chief god of the city of Babylon, who had also been a god of thunderstorms. His elevation was justified by a story that demonstrated what a fearful deity a storm god is. It was said that long ago he had fought an army led by the monster of primeval chaos, Tiamat. Leaping into his storm-chariot, 'the unequalled and terrible tempest', Marduk fired off his lightning and let loose his arsenal of winds including 'the storm, the parching blast . . . the typhoon, the wind incomparable'.[5] Tiamat was overwhelmed, and Marduk cleft open her bowels and pierced her heart. Following her defeat, according to the legend, the monster's consort, Kingu, was taken captive and his blood used to create human beings.

Adad, a god of the Babylonians and the Assyrians, had like Ninurta an ambivalent persona as both destroyer and giver of life. His storms brought darkness, want and death, but his rains enabled the land to produce food. The same was true of Ninhar, the city god of Kiabrig, near Ur. Represented as a roaring bull, he was the deity of thunder and of the rainstorms that made the desert green. For the Canaanites the king of the gods was the storm deity Baal, or 'He Who Rides on the Clouds'. Baal also represented life and fertility and fought a never-ending war with the god of death and sterility. When he was in the ascendant, crops grew. If he was doing badly, there was drought.

Baal with Thunderbolt, carving from around the 15th century BC.

Not surprisingly, earthly potentates tried to exploit the awe that storms inspired. The king of the Hittites was regarded as the earthly representative of their storm god, Tarhun, and King Idrimi of Alalakh, in modern-day Turkey, described himself as a 'servant of the storm god', whom he proclaimed 'the lord of heaven and earth'.[6] For others, though, representing the storm god was not enough. Salmoneus, the king of Elis in Greece, imitated thunder by dragging bronze kettles behind his chariot or driving over a bronze bridge while he hurled burning torches in an attempt to mimic lightning. He declared he was Zeus and demanded that sacrifices be made to him. Similar delusions gripped one of the ancient kings of Alba Longa in Italy in the eighth century BC. To prove he was a god, equal or superior to Jupiter, he built machines that simulated the sound of thunder or the flash of lightning, and when real storms broke he would make his soldiers bang their swords on their shields to try to drown out the thunder. Jupiter eventually appears to have tired of these antics, because the story goes that, during a fierce storm, the king was killed and his palace destroyed by a thunderbolt.

It was only to be expected that the chilly regions of northern Europe would also spawn a celebrated storm god. Thor, red-bearded and prodigiously strong, was worshipped by all the early Germanic peoples. His name was their word for thunder, and in countries such as Iceland he was honoured more than any other deity. Whenever there was a fight, he was the gods' champion, an implacable foe to their enemies but a friend to mankind. The eleventh-century German chronicler Adam of Bremen recorded the benefits Thor brought to humanity because he controlled not only thunder, lightning, wind and rain, but 'fine weather and crops'.[7] Like Jupiter Thor had a day named after him: Thursday. He was said to travel in a chariot drawn by goats, and thunder was believed to be the noise made by this conveyance. The thunderbolt was his main weapon, represented by Mjölnir, a magic short-handled hammer forged by dwarves, which had a number of wondrous attributes, including the ability to return to its thrower like a boomerang. The story goes that a giant named Thrym once stole the hammer and demanded the hand of a goddess in marriage as the price of its return. Thor went to the nuptials disguised as the bride, and evaded detection in spite of eating an ox and eight salmon. He then grabbed the hammer and used it to slaughter Thrym and some other giants. Just as it had been for the ancient Greeks, for northern Europeans the oak was closely identified with the thunder god, whether that was Thor or Perun, his Slavic equivalent.

Perun was similar to Thor in a number of ways. He too had an axe that would return to his hand after he threw it, and his chariot was drawn across the sky by an enormous billy goat. The Slavic god was much feared because he used thunder and lightning to strike down people who had offended him, but he also promoted fertility and in spring his thunderbolts awoke the earth from the death-like sleep of winter. There was a belief that the darkness held the sun prisoner in a cell, the door of which could be opened only by Perun's lightning. He played an important role in battles, bestowing victory on those soldiers he favoured. It is said that at Novgorod, a fire of oakwood burned day and night in his honour, and that if it ever went out, the negligent

Mårten Eskil Winge, *Thor's Fight with the Giants*, 1872, oil on canvas.

attendants would be executed. According to the sixth-century Byzantine historian Procopius, the Slavs regarded him as their chief god, the 'lord of all things'.[8] Sacrifices of cockerels, oxen, other animals and even humans were made to him. A spring festival was held during which maidens would dance themselves to death in his honour (this became the inspiration for Igor Stravinsky's famous ballet *The Rite of Spring*). In AD 988 Vladimir I, grand prince of Kiev, agreed to bring Christianity to his realms as part of a peace deal with the Byzantine emperor. He ordered the destruction of all pagan gods, then took the statue of Perun that stood outside his palace, tied it to a horse and dragged it through the streets before throwing it in the river. Even so, Christian missionaries found it hard to stamp out the worship of Perun, and it persisted long afterwards. In some areas he became amalgamated with the Old Testament prophet Elijah, who shared many of his attributes, such as the power to call down rain or fire from heaven.

Perun's equivalent in Lithuania, Perkuns or Perkunas, also had everlasting fires of wood from sacred oak trees kept burning in his honour. He was sometimes believed to have played a crucial role in creation as the divine smith who moulded the world. One of his duties was to act as god of justice: he killed the goddess of the sea because of her love affair with a mortal, and broke the face of the moon god for kidnapping the planet Venus. He relentlessly pursued the Devil, who would sometimes hide in a tree until Perkunas zapped it with lightning. Even after the arrival of Christianity, farmers would sacrifice animals to him in times of drought, drink beer in his honour or dance around bonfires in the hope of stimulating rain. The Finns believed a thunderstorm was a sign that their thunder god, Ukko, was making love to his wife, while in the British Isles and Gaul, the Celts worshipped the storm god Taranis. His sacrificial victims, human or animal, were said to have been placed inside great wickerwork images, then burned – a ritual that is echoed in the cult 1970s British horror film *The Wicker Man*, in which a devoutly Christian police officer is shocked to find himself on a remote island where paganism flourishes and couples copulate openly in the fields. He ends up being sacrificed inside a modern Wicker Man effigy.

Far from the shores of Europe, in India too the storm deity, Indra, was king of the gods in early Hindu mythology. He rode a huge white elephant with four tusks, and was a great warrior, defeating the enemies of the gods in battle. Diti, the mother of demons, decided to produce a son more powerful than Indra. Her husband said it could be done, but that she would have to remain pregnant for a century, during which time she must scrupulously observe all religious rites. Diti agreed to comply with these onerous conditions, but while she was asleep Indra took his sword and crept into her womb through her nostril. He cut her child into 49 pieces. Eventually each piece became a

Indra riding on his four-tusked elephant, *c.* 1830, watercolour.

marut, a kind of assistant storm god who served Indra. Regular sacrifices of food had to be made to keep Indra happy, and a Hindu sect, the Vallabhacharya, who worship the great god Krishna, told a story of what happened when he got jealous. Some cowherds had been preparing to worship Indra when Krishna persuaded them instead to pay homage to Mount Govardhana, of which he became the spirit. In his fury, Indra sent a dreadful week-long storm. To protect the cowherds and their cattle, Krishna had to lift the mountain like an umbrella over them.

A number of storm gods are associated with chaos and disruption. One was the ancient Egyptian deity Seth. With a canine body, slanting eyes, a long, pointed snout and a forked tail, he was also the god of warfare and disorder – embodying the creative value of shaking up the status quo. The Japanese storm god Susano-o-no-Mikoto, meaning 'impetuous male', could also be antisocial, damaging rice fields and throwing a flayed horse into his sister's weaving hall. Expelled from heaven, he did a good deed on earth, slaying an eight-headed dragon that had been terrorizing the countryside. The Chinese god of thunder was Lei Gong, a fearsome blue creature with claws and bat's wings, clad only in a loincloth. He made the sound of thunder with a drum and a mallet, and carried a chisel to punish wrongdoers. His retributive duties led to people doing him special honour in the hope that he would harm their enemies. Although he had a foul temper, there was also a kinder side to Lei Gong. One day a young man was collecting firewood and medicinal herbs on a mountain side when he heard a loud clap of thunder. Looking up, he saw Lei Gong trapped in a split in a tree. The thunder god explained he had been trying to break the tree when he got stuck, and offered the boy a reward if he would help to free him. Once he was out, he told the boy that if ever he was in trouble, he could count on his godly assistance, though he warned that this should not be invoked for anything trivial. The boy did good works, curing the sick, but he also fell foul of a local official who had him arrested. Only then did the boy call on Lei Gong, who promptly unleashed a thunderclap of such ferocity that the official was afraid for his life, and released his prisoner. Lei Gong's wife was the lightning goddess and carried a brass mirror, which she flashed to generate lightning. He also had assistants, such as Yuzi, who caused downpours by dipping his sword in a pot, and Feng Bo, who released roaring winds from a goatskin bag. But weather gods in China were not always treated with respect. In 1888 Canton had been suffering incessant rain, and government officials prayed to a god named Lung-wong to halt it. When he failed to deliver, they locked up his effigy for five days, releasing it only when the rain stopped.

When Europeans first explored the Americas, they ran into a dreadful new kind of storm. One sixteenth-century writer described how such a tempest came

> with great violence, as if it wanted to split heaven and earth apart . . . it thundered so cruelly with cracks and crashes, and the lightning flashes came so quickly after one another that the sky seemed to be completely full of fire. Soon after that a thick and dreadful darkness came to the day which was even darker than any night could ever be.[9]

The storm was so fierce 'it ripped many large trees out of the earth by the roots and threw them over', while many houses and villages were also destroyed. Columbus was fairly lucky with the weather on his first three voyages, though on his second he lost three ships in what was probably a waterspout. But he had heard the local Taino people talking about storms of almost unimaginable power, using the name 'Huracan'. On his fourth expedition in 1502, he noted the ominous swell on the sea and threatening clouds above, which they said presaged such an event, and took refuge in a harbour. Another Spanish fleet had 25 out of 30 ships sunk in the fearsome storm that developed. Huracan, or Jurakan, the Spanish explorers discovered, was an evil god of winds and destruction, and the native people would shout and beat drums to try to keep him away. The Taino, who inhabited Cuba, Hispaniola and other parts of the Caribbean when the Spaniards arrived (within half a century they had been almost wiped out through

Christopher Columbus's first landing in the New World, as imagined in a chromolithograph from 1893.

ill-treatment, starvation and disease) believed he was the son of the goddess Atabei, who created the earth and the sky. He had a brother who carried on the good work, making the sun and moon, and populating the earth with plants and animals, but this made Huracan jealous and he began to tear up the world with dreadful winds. Likenesses show him with two distinctive arms spiralling out from his sides, leading to speculation that the Taino had deduced, perhaps from the configuration of damage on the ground, the vortex pattern of hurricanes which modern humans began to detect only hundreds of years later.

On the mainland the Aztecs were a formidable people who conquered a huge empire in central and southern Mexico, but they were very much afraid of a god named Tlaloc, and sacrificed children to him. He could hurl lightning, send or deny rain and unleash devastating hurricanes. Wearing a peculiar mask, with big round eyes and long fangs, he was also able to deploy illnesses such as leprosy and dropsy. People who died from these diseases

Willem van de Velde the Younger, a ship on the high seas caught in a squall, *The Gust*, *c.* 1680, oil on canvas.

or were struck by lightning were believed to enjoy a blissful life in paradise after their deaths. The Maya, who had a huge empire in Mexico, Guatemala and Belize, also made human sacrifices to a similar god named Chac at Chichen Itza. North American native peoples, too, had storm gods. The Maidu of California believed that a Great Man created the world and all its people, and that when lightning strikes, it means the Great Man is making a visit from heaven. Many North American tribes revered the thunderbird, from the beak of which lightning flashed, while the beating of its winds generated the growl of thunder. Thanks to this powerful spirit, the earth was watered

Monolith of the Aztec god Tlaloc, who could hurl lightning and unleash hurricanes, from the National Museum of Anthropology, Mexico City.

The pyramid of Kukulkan, Chichen Itza, Mexico, built *c.* 9th–12th century.

and made fertile. Evidence of similar spirits has been found in Europe, Asia and Africa. In parts of Nigeria today, the Yoruba people still revere the storm god Shango, who carries a double-headed axe and is accompanied by a ram. At his annual festival his priest dances himself into a state of deep trance, expressing Shango's wrath through the lightning movement of his arms.

Further evidence that the storm god is a truly worldwide phenomenon comes in a colourful Maori myth from New Zealand. It tells how the sky god Rangi made love so endlessly to the earth goddess Papa that their children could never escape from their mother's womb. Eventually, after a number of attempts, one of their offspring, the god of forests, managed to prise the affectionate couple apart. This upset the storm god Tāwhirimātea because he had been quite happy in the womb. So he sent fierce squalls, whirlwinds, thunderstorms and hurricanes, blowing down the forest and forcing the fish that lived in its branches to take refuge in the sea. Only Tūmatauenga, the god of fierce human beings, stood up to him, and ate all his brothers and sisters, leaving him to fight Tāwhirimātea alone. But despite a long struggle, which caused a flood big enough to create the

Painted carving of a thunderbird, from British Columbia.

Ceremonial staff with a carving of Shango, Nigeria, *c.* 1900.

Pacific Ocean, he was unable to defeat the storm god, and had to let him live on as an enemy of man on land and sea.

Storms also play an important role in the Bible. Sometimes God employs them as weapons. Just as in the *Epic of Gilgamesh*, God uses a fierce and lengthy rainstorm, opening 'the windows of heaven', to create the great flood that wipes out the whole of humanity apart from Noah and his family in their Ark (Genesis 7:11). Later in the Old Testament the Almighty sends a series of plagues against the Egyptians to make them release the Israelites from their enslavement. Along with the frogs, flies, boils, locusts, cattle sickness and so on is a fearful hailstorm:

> There was hail, and fire mingled with the hail, very grievous, such as there was none like it in all the land of Egypt since it became a nation. And the hail smote throughout all the land of Egypt all that was in the field, both man and beast; and

> the hail smote every herb of the field, and brake every tree of the field. (Exodus 9:24–5)

God also used storms to attract and keep attention. When He was summoning the Israelites to Mount Sinai to receive the Ten Commandments, 'there were thunders and lightnings, and a thick cloud upon the mount' (Exodus 19:16). The people trembled and presumably became a suitably receptive audience. Storms also feature in other memorable Old Testament stories, such as that of Jonah and the whale. The Almighty instructed the prophet to go to the city of Nineveh and tell the people there how wicked they were. Jonah was not keen, and instead quickly hopped on a ship going in the opposite direction, but God 'sent out a great wind into the sea', causing 'a mighty tempest', so the mariners thought the vessel was going to be wrecked. They cast lots to decide who was to blame for the storm, and the verdict was – Jonah. Rather nobly he owned up to having upset God and told the boat's crew to throw him in the sea. At first they were reluctant, but as soon as they tossed him overboard 'the sea ceased from her raging'. Famously Jonah was then swallowed by 'a great fish', in whose belly he spent three days and three nights before being vomited out, by God's orders, on dry

'And Moses stretched forth his rod toward heaven, and the Lord sent thunder and hail' (Exodus 9:23), German woodcut from 1877.

land (Jonah 1:4–17). Jonah then goes to Nineveh and tells the inhabitants of the evil of their ways. Far from beating him up, they put on sackcloth and ashes and repent.

Storms also feature at important moments in the New Testament. According to St Matthew, just after Jesus had delivered the memorable line 'The foxes have holes, and the birds of the air have nests; but the Son of Man hath not where to lay his head', he and his disciples boarded a ship to cross the Sea of Galilee. Soon the weather turned nasty. In Matthew's words: 'there arose a great tempest in the sea, insomuch that the ship was covered with the waves'. Jesus is sleeping serenely, so the disciples wake him, crying: 'Master, carest thou not that we perish?' Christ rebukes them: 'Why are ye fearful, O ye of little faith?' Then he speaks to the sea: 'Peace, be still. And the wind ceased, and there was a great calm.' A similar episode occurs after Christ's crucifixion and resurrection, as his apostles set out to spread his word. On St Paul's final voyage, they encountered 'a tempestuous wind' and were soon in fear for their lives: 'when neither sun nor stars in many days appeared, and no small

Fresco showing Jonah and the whale, painted *c.* 1480 by Albertus Pictor in the vaults of Härkeberga Church, Uppsala, Sweden.

Jonah and the whale, stained glass window, Church of St-Aignan, Chartres, 19th century.

tempest lay on us, all hope that we should be saved was then taken away.' Then Paul told them the angel of God had appeared to him, telling him not to be afraid, and declared: 'be of good cheer: for there shall be no loss of any man's life among you.' And, indeed, after drifting for fourteen days, the vessel was wrecked on the coast of Malta, and all the disciples survived.[10]

The high status of storm gods illustrates how desperately human beings want to take some kind of control over tempests, so it is not surprising that they are at the centre of many superstitions. Perhaps because of their association with witches, cats are sometimes seen as having a connection with storms. A rhyme from 1773 tells how feline behaviour can predict wild weather:

> Against the times of snow or hail,
> Or boist'rous windy storms;
> She frisks about and wags her tail,
> And many tricks performs.[11]

Sailors, in particular, would become anxious if a cat grew too lively, saying it had 'a gale of wind' in its tail. They also believed that throwing a cat overboard was a sure recipe for generating a storm. But dogs were also sometimes linked with tempests. One August Sunday in 1577 a violent thunderstorm damaged the villages of Blythburgh and Bungay in Suffolk, and a contemporary report claimed that a black dog of 'horrible shape', looking like the Devil and accompanied by 'fearful flashes of fire', was seen rushing around, killing or injuring a number of people.[12] Then there was hair. The Tlingit Indians of Alaska thought storms could be caused by a girl rashly combing her hair outdoors, while in the Scottish Highlands it was said that no woman should ever comb her hair at night if she had a brother at sea. The Romans believed that no one on board ship should ever cut his hair or nails *except* in a storm, because by then there was no point in trying to stop the tempest. Maoris in New Zealand took elaborate precautions during haircuts. Spells against thunder and lightning were said, and afterwards both the person whose hair was being cut and the cutter were put under various taboos, such as not being able to touch food, associate with other people or

Ludolf Bakhuizen, *Christ in the Storm on the Sea of Galilee*, 1695, oil on canvas.

resume their normal work for some time afterwards. In the Tyrol witches were supposed to be able to use cut hair to summon up hail or thunderstorms.

As we have seen, the ancient Greeks revered trees struck by lightning, and in other cultures too they were regarded as having magical powers. In Central Europe a strike on a tree was believed to create a 'thunder-besom' – a growth that looks a bit like a nest.[13] The superstition was that if a family burned it in the hearth, it would protect the house from being struck. On the other hand, some Saxon peasants feared that burning wood from trees that had been struck might actually set the house on fire. The same view was taken by the Tsonga people of southern Africa, while one tribe in British Columbia would be sure to use fire arrows made from the wood of a tree that had been struck when they wanted to set alight the houses of their enemies.

The Nootka people of British Columbia believed twins had power over storms, while among the Bantu tribes of Delagoa Bay in Mozambique, it was the mothers of twins. If storms had not been providing enough rain, the tribe's women would strip naked and then put on girdles and headdresses of grass or short petticoats made of leaves. Shouting and singing ribald songs, they would go around to each well to clear out mud. Next, carrying pitchers of water, they would move on to the house of a woman who had borne twins, and drench her. Sometimes they would visit the sacred grove and pour water on the graves of twins. If this failed, they would take the body of a twin and rebury it closer to water. At times humans, particularly sailors, need winds rather than rain. Just as the Chinese believed that there was a god who kept winds in a bag, so the ancient Greeks told how Ulysses received them in a leather pouch from Aeolus, lord of the winds. Old women in Shetland and wizards in Finland would sell wind tied up in knots to mariners. The Finnish sailors got a moderate wind if they undid the first knot, a gale if they progressed to the second, while the third unleashed a hurricane.

When it comes to quietening a storm, that Roman of many talents Pliny the Elder wrote in his *Natural History* that it was widely believed that if a menstruating woman stripped naked, she could scare away 'hailstorms, whirlwinds, and lightning'.[14] In Greenland it was women who are about to give birth, or who have just given birth, who were thought to have the power of stilling a storm. All they had to do was go outside and fill their mouths with air, then come back in the house and blow it out. Eskimos had a number of expedients for halting storms when they had been raging for too long and food was short. One was to take a long whip made of seaweed to the beach and strike out in the direction of the wind, shouting: 'Enough!'[15] Another was to light a fire on the shore, around which men would gather and chant. In a coaxing voice, an old man would invite the demon of the wind to come and warm himself. Once they believed he had arrived, the Eskimos would throw water on the fire and then fire arrows at where the flames had been. They believed the demon would not stay somewhere he had been so badly treated. In Alaska Eskimo women would drive away the spirit of the wind by making passes in the air with clubs and knives, while men fired rifles at him. This idea of fighting a storm seemed to be widely practised. Swords were used in Borneo and Sumatra, where the local rajah took the lead in the attack, while Bedouin in East Africa relied on daggers. If the wind blew down their huts, the Payaguas of Paraguay would beat it with their fists, or threaten it with firebrands. The Khoikhoi of southern Africa tried a different approach. They would take one of their heaviest animal skins and hang it on a pole, believing that in trying to blow it down, the wind would exhaust itself.

When Sir James George Frazer, considered by many to be the father of modern anthropology, published his revolutionary study of magic and religion, *The Golden Bough*, in 1890, he wrote that in many villages in Provence there was still a belief that some priests had the power to prevent storms. When a new priest took over, local people would be keen to test his abilities, so at the first sign of a tempest they would ask him to exorcize the clouds. If this worked, he won the respect of his flock. Frazer added that anti-storm rituals were widespread in Roman Catholic countries. On the day before Easter Sunday, it was customary to put out all the lights in the churches and then start a new fire to light the great Easter candle. Local people would bring sticks of oak and other wood, which they would char in the fire, then take home to burn while they prayed that God would protect the house from lightning and hail. Some of the sticks were kept to insert into the roof if a heavy thunderstorm did strike, while others were put in the fields to stop crops being beaten down by hail. Frazer wrote that Midsummer's Eve was also an important time for anti-storm rituals. In Bohemia the people would fell a fir tree and decorate it with wreaths and red ribbons, then at nightfall set it on fire. People took home the singed wreaths and during thunderstorms would try to protect their homes by burning them in the hearth while they said a prayer; in Poitou they lit bonfires and used the ashes for protection. Around Christmas in many parts of Europe it was the custom for households to burn a big, hard log, often of oak, known as the Yule log. In some villages in Westphalia, this was fitted into the hearth, then removed as soon as it was slightly charred. Afterwards it was stored away to be put back on the fire whenever a thunderstorm broke, because people believed that lightning would not strike a house in which a Yule log was smouldering.

2 Nature

The ancient Greeks may have had their storm gods, but they are generally regarded as the first people to try to find some kind of scientific explanation of how the weather works, rather than just putting it down to divine intervention. During the seventh century BC Thales of Miletus, often regarded as the first Greek philosopher, developed a theory that the weather was influenced by the movement of planets and stars. The story goes that this gave him such wonderful powers of prediction that one winter he was able to foresee that the following year's olive harvest would be a bumper one, so he booked up all the local olive presses in advance. Then when growers did indeed have olives coming out of their ears, Thales was able to name his price for renting out the presses and made a fortune. A couple of hundred years after Thales, another Greek philosopher, named Anaxagoras, tried to explain the apparently bizarre phenomenon that although hail was ice, hailstorms often happened when it was hot. He came up with the idea that hail formed when heat from the earth pushed clouds up to a cold belt high in the atmosphere. Anaxagoras also had a theory about lightning, saying it resulted from sparks escaping from a zone of fiery 'aether' above the cold belt. When these sparks collided with clouds lower down, they split them apart, producing thunder and lightning. Then around 340 BC, the great philosopher Aristotle produced his *Meteorologica*, the first attempt to formulate a comprehensive science of weather. He got plenty of things wrong. For example, he rejected Anaxagoras' belief that hailstones originated in the

Lightning during a storm in Hanover, Germany, 2013.

upper air, maintaining instead that 'they froze close to the earth', but other features he understood pretty well, writing, for example, that thanks to the sun, 'the finest and sweetest water is every day carried up and is dissolved into vapour and rises to the upper region, where it is condensed again by the cold and so returns to the earth.'[1] Aristotle is still regarded as the father of meteorology and his principles were not seriously challenged for the next couple of millennia, until the seventeenth century when new scientific instruments such as barometers and thermometers appeared, allowing scientists to make the first precise measurements of weather phenomena.

An important figure who advanced our understanding of storms was the astronomer Edmond Halley, after whom the famous comet was named. In 1686 he produced the first meteorological map of trade winds, and advanced the theory that winds, those crucial ingredients of storms, are caused when air heated by the sun rises and other air rushes in to fill the gap. It was an important step forward, but plenty of questions remained. In the nineteenth century the eminent American meteorologist James Pollard Espy, dubbed the 'Storm King', argued that air rushed straight from all directions towards the centre of a tempest because this was the area of lowest pressure, but a self-taught meteorologist from Cromwell, Connecticut, who went on to become the first president of the American Association for the Advancement of Science, reached a different conclusion. Carefully observing the positions of trees that had been felled by a great gale in New England in 1821, William Redfield maintained that the wind must have taken a circular path. A similar idea had already been suggested by Colonel James Capper of the British East India Company. His observation of tropical storms off the coast of the subcontinent made him think they were 'whirlwinds'.[2]

Further advances in understanding storms came at the end of the First World War, from the Norwegian physicist Vilhelm Bjerknes and his son Jacob. They hit upon the crucial role played by the collision of warm and cold masses of air. These masses are sometimes thousands of miles across, and have sat for long enough over a region to pick up its characteristics of temperature and humidity, which are spread fairly uniformly through them. Influenced by the titanic clashes of armies during the war, the Bjerknes gave the name 'fronts' to the places where these masses meet.

Today many questions remain about how exactly storms are created, and although some of the most powerful supercomputers in the world are enlisted to predict where and when they will appear, forecasters often get it wrong. Modern theories are based on the idea that storms need three basic ingredients – energy, moisture and unstable air. The energy can come from the temperature differences between two neighbouring air masses or from the sun heating up the earth's surface. The warmer the surface, the more energy it can supply. Moisture not only provides rain, snow or hail but also furnishes the means of transferring energy from the earth's surface to the higher reaches of the atmosphere through water vapour, produced when the sun's heat evaporates water. This change of state from liquid to gas means the vapour absorbs energy in the form of latent heat. As the vapour rises, it condenses back into water, releasing the latent heat as energy, which helps to fuel the storm. The final ingredient, unstable air, is generated when the air surrounding a storm cloud is significantly colder than the air lower down. Warm air is lighter than cold air, so this enables the storm cloud to keep on rising. The greater the temperature difference, the higher, faster and stronger the storm will grow.

Storms are generated by both global and local factors. On the global scale, the sun's rays do not

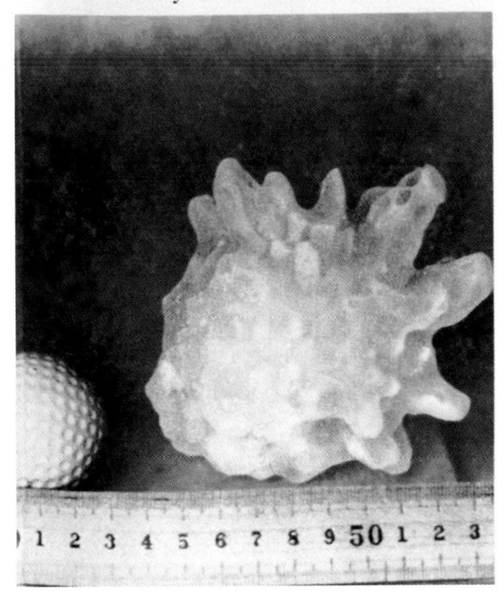

Hailstones can reach lethal sizes. This one is said to have fallen on Washington, DC, in 1953.

warm all parts of the earth equally, hitting the equator much more directly than the poles, making equatorial regions hotter. Indeed, the weather can be seen as a machine that is constantly trying to even out these differences in temperature, but which never quite succeeds. A storm is one of nature's most effective ways of moving heat around. As Halley had suggested, warm air nearer the equator rises and moves towards the poles, while cold air moves away from the poles to replace it. However, Redfield and Capper were right in their claim that it does not flow directly from high- to low-pressure areas. Instead the rotation of the earth skews the direction of winds. As our planet spins from west to east, it curves their path to the right in the northern hemisphere and to the left in the southern.

As for the clashes of giant masses of air that the Bjerknes uncovered, their effects can be observed over the continental United States where cold air coming down from Canada collides with warm air rising up from the Gulf of Mexico. This generates plenty of extreme weather, with perhaps 10,000 severe thunderstorms every year. Because cold air is heavier than warm air, a cold front hugs the ground. When it runs into a warm front, it behaves like a wedge, driving the hot air upwards until it cools, bringing heavy rain. If a warm front is advancing, it rises over colder air. Then it cools, and the water vapour in it condenses to form cloud and rain. A cold front travels faster, so it can also catch up with a warm front and push underneath it, again bringing heavy rain.

In the UK, too, some of the country's most severe weather comes when cold, dry air from the north runs into warm, humid air from the south. Polar or Arctic air masses are cold, while tropical ones are warm. Maritime masses form over water and are more moist. Continental masses form over land, and tend to be drier. The polar maritime air masses that begin their journey over northern Canada or Greenland start out cold and dry. Then, as they pass over the Atlantic, they pick up heat and moisture from below. As they get hotter, they rise, which makes them cool again, so by the time they reach the UK, they can bring rain, hail, sleet or snow. Polar continental air masses come from the North Pole, travelling mainly overland, sometimes across Siberia and Eastern Europe and sometimes over Scandinavia. They remain fairly dry until they reach the North Sea, where they can pick up enough moisture to deliver snowstorms to the east coasts of England and Scotland. One dumped 40 cm (16 in.) of snow on Northumberland in November 2010.

A storm can also be generated by more local conditions. An area that is paved or rocky, for example, heats up much more quickly than the surrounding land, while a black surface, such as a tarmac road, can get up to 40°C warmer than a white one. On a bigger scale, cities often have microclimates that mean they are hotter than the land around them. If the difference in temperature is big enough, this can soon generate an updraught sufficiently powerful to form a cumulonimbus cloud. These monsters can tower 16 km (10 miles) into the air. They are often shaped like an anvil, with their tops flattened by the strong winds found high up in the atmosphere, but their dark, glowering bases may be only a few hundred metres above the earth. Their appearance heralds storms, often accompanied by teeming rain, hail or lightning. Inside the cloud, the powerful updraughts carry droplets of water so high that they begin to freeze, forming little ice crystals. These then start to drop, dragging air with them and creating a downdraught. As they fall, they collide with other water droplets or ice crystals still being lifted by updraughts. These collisions, which happen millions of times each second, generate an electrical charge. Lighter, positively charged crystals gather in the top of the cloud, while heavier, negatively charged particles gravitate to the bottom.

This causes a positive charge to build up on the ground beneath the cloud, concentrating around anything that sticks up, such as mountains, buildings, trees or people. When the charge coming up from these points connects with the charge coming down from the cloud, lightning strikes them. A second stroke of lightning then returns from the ground to the clouds along the same channel. It is the return stroke that we see. Its heat raises the temperature of the surrounding air to about 27,000°C (48,632°F) – around five times hotter than the surface of the sun.

Mammatus clouds, often seen when thunderstorms are close, photographed here over Squaw Valley Ski Resort in California, 2010.

Because this all happens so quickly, with lightning travelling at about 121 km per second (75 mps), the dramatically heated air has no time to expand, and its pressure races up to perhaps 100 times its normal level. This causes an explosion we hear as thunder. Because light travels so much faster than sound, we see the lightning well before we hear the thunder. So if the lightning strike is a mile away, we see it five seconds before we hear the accompanying thunder. In fact, only about one lightning stroke in five hits the ground, with the others going from cloud to cloud. Inside the thunderhead, the water droplets or ice crystals eventually get heavy enough to resist the updraughts and fall to earth as rain or hail. Once the updraughts have been completely replaced by downdraughts, this cuts off the supply of heat and moisture, and the storm begins to die away. Most thunderstorms last no longer than half an hour and cover only a few square kilometres, but it is estimated that lightning strikes somewhere on the earth more than 40 times every second, or nearly 1.3 billion times a year, and that it kills up to 24,000 people every year. The average lightning flash would power a 100-watt light bulb for three months, and during a major thunderstorm an inch of rain might fall. And far from lightning never striking the same place twice, a structure as tall as the Empire State Building can be hit more than 40 times in a single day.

Since they depend on air being heated quickly, thunderstorms are more common over land than sea, and are most frequent in hot and humid tropical areas. In more temperate zones they usually happen in the afternoon or evening, when the sun has had longer to heat up the ground. The world record for thunderstorms was set by Bogor on the Indonesian island of Java, where there was an average of 322 each year between 1916 and 1919. Many tropical regions can expect thunderstorms on 200 days out of a year, while for places in the southeast of England the number is more like twenty, and in parts of Scotland it is fewer than five.

In 1959 a U.S. Marine Corps pilot got a unique insight into the chaos inside a thundercloud when he had to bail out of his

Mount Salak, in the Bogor region of Java, Indonesia, which holds the world record for frequency of thunderstorms.

aircraft and fell right through one. Lt-Col. William Rankin is the only person known to have survived such an experience. He said the violent up- and downdraughts inside it bounced him like a ping-pong ball on a jet of water for three-quarters of an hour. He was surrounded by flashing lightning and deafened by thunderclaps, while the rain was so torrential that he feared he was going to 'drown in midair'.[3]

So Anaxagoras was right when he suggested that hail is produced when warm air is driven high into the atmosphere, where it is colder. When it gets up there, the water vapour inside it condenses into little drops of water or ice crystals. First these form clouds; then, if conditions are right, they fall as rain, snow or hail. The hailstones that often accompany thunderstorms first form around a nucleus, generally a tiny particle of dust or salt, though one that came down in Oklahoma in 1975 had a little wasp at its heart. In the turbulent interior of a cumulonimbus cloud, a stone may fall for hundreds of metres before being caught in one of its powerful updraughts and carried back up to where temperatures might be as low as −20°C, so that it acquires another coat of ice. This process can happen a number of times, and occasionally stones have been found with more than twenty

layers. When they become too heavy for the updraughts to support them any longer, they fall as hail. The smallest stones have a diameter of about 5 mm but in Britain, some as heavy as 280 grams have been reported, though even they were dwarfed by the 1-kg monster that descended on Bangladesh in 1986. When the British moved into India, they were astonished by the dimensions of the hailstones. A doctor named A. Turnbull Christie recorded that some the size of walnuts had fallen on Tiruchirappalli in 1805, while at Darwar in 1825 they were as big as 'a pigeon's egg'. There was even a story, related the doctor, that one the size of an elephant had landed at Seringapatam during the late eighteenth century, though he added that we should not 'believe this to the letter – we must make some allowance for oriental exaggeration'. Still, the hailstones had their compensations. After one storm at Hyderabad in 1823 'a sufficient quantity were collected by the servants of a military mess to cool the wine for several days.'[4]

Hailstorms tend to last for no longer than fifteen minutes, which is fortunate, considering that the bigger stones are hitting the ground at perhaps 160 km/h (100 mph). Not surprisingly they can do a lot of damage, bringing down power lines, damaging trees and killing animals and people. In 1953 an American biologist witnessed the devastation caused by one that swept across Alberta in Canada: 'Grasses and herbs were shredded beyond recognition and beaten into the earth. Trees and shrubs were stripped of all leaves and small branches, and the bark on one side of the larger trees had been torn away or deeply gouged.'[5] Birds were killed by the thousand, including an estimated 36,000 ducks. Nowadays it is reckoned that hail destroys up to 6 per cent of the total crop yield in the High Plains of the United States, and across the country the cost to agricultural production is put at about $1.3 billion a year, with damage to property estimated at around $1 billion. Hailstones can also bring down aircraft. In 1977 a Southern Airways DC-9 flying from Huntsville, Alabama, to Atlanta, Georgia, ran into a hailstorm. It had its windscreen smashed and so much water and ice got into its engines they stopped working. The crew tried to make an

People stranded on a bridge outside Bangkok, Thailand, surrounded by floodwaters during the 2011 floods.

emergency landing on a stretch of road, but one wing clipped a petrol station, causing the aircraft to burst into flames, killing 63 people on board and nine on the ground.

Hail's close relatives – rain and snow – can also feature in fearsome storms. The record for the heaviest rainfall in a single day was 1,825 mm (71¾ in.) on 7–8 January 1966 on the island of Réunion, while on Guadeloupe on 26 November 1970, 38 mm (1½ in.) fell in just one minute. (The most rain to fall in 24 hours in the UK was 316 mm (12½ in.) at Seathwaite in the Lake District on 19 November 2009, while more than 32 mm descended on Preston in just five minutes on 10 August 1893.) Devastating rainstorms are probably encountered more frequently than any other type of tempest. Just take the year 2011. On 12 January parts of Brazil experienced more than a month's rainfall in 24 hours, resulting in floods in São Paulo State that killed at least 24 people, and landslides in Rio de Janeiro State that cost 900 lives. At the same time, in Australia, parts of Queensland were experiencing the worst inundations

A police car drives through storm water during a monsoon, Kerala, India.

in 40 years, with at least twenty people killed when a 7.9-m (26-ft) flash flood swept through the town of Toowoomba. On 13 January the Sri Lankan government announced that 40 people had perished and 300,000 had been driven from their homes after heavy rain caused floods and landslides. Next it was the turn of South Africa, where the death toll from rainstorms and flooding reached 100 by the end of the month. March and April brought downpours that devastated southern Thailand, while in May rainstorms in China and Malaysia caused landslides that killed about twenty people in each country. The following month, heavy rain brought deadly floods to three different parts of China, while Haiti also suffered heavy casualties. In July it was the turn of South Korea, then, in August, Uganda. September's monsoon rains killed at least 80 people in India, plus 400 in Pakistan, where 665,000 homes were damaged or destroyed – and so it went on.

We get snow when it is cold enough and there is sufficient moisture around for ice crystals to form in the atmosphere. As these crystals collide, they form snowflakes. If enough crystals join together, they become sufficiently heavy to reach the ground. When the temperature is 2°C or below, we have snow. If it is higher the snowflakes will melt as they come down, and arrive as sleet. A rancher at Fort Keogh in Montana in 1887 said he had seen snowflakes 'larger than milk pans' and measured one at 38 cm (15 in.) wide.[6] If that is true, they would be the biggest ever recorded, but there was no one to corroborate the claim. However, in 1988 a weather observer for Britain's Royal Meteorological Society said he himself had seen flakes up to 7.6 cm (3 in.) across in Vancouver, and that he had found a dozen credible reports of specimens up to 15.2 cm (6 in.) wide. The heaviest known snowfall in a day was 193 cm (76 in.) in April 1921 at Silver Lake, Colorado, while nearly 28.3 m (93 ft) fell in a year at Mount Rainier in Washington State from 19 February 1971 to 18 February 1972. The most devastating snowstorms, though, are blizzards, where strong winds drive the snow, often whipping up deep drifts. America's National Weather Service defines a blizzard as a storm of at least three hours' duration, with large amounts of snow and winds with speeds of more than 35 mph, like the kind that hit the Northwest Plains of the United States on 12 January 1888.

The weather had been unusually mild, but the temperature fell nearly 41°C (75°F) in just a few hours. Suddenly the region found itself in the grip of what became known as the 'Schoolhouse Blizzard' because so many of its 235 victims were children on their way home from school. At Mira Valley, Nebraska, pupils were sheltering with their teacher, Minnie Freeman, when the wind tore off the roof of their one-roomed schoolhouse. She felt she had no option but to try and get the children to somewhere safe. According to some accounts, she used a clothes line to tie them to her and lead them to a farmhouse. Every child survived, and her heroism is commemorated on a mural in the state's Capitol Building. Another Nebraska teacher, Loie Royce, decided to take her three pupils to her home when the schoolhouse ran out of fuel at three o' clock in the afternoon. It was a journey of less than 82 m (90 yds), but they got lost in the blinding snow almost immediately, and as the group wandered helplessly, the three children all died of hypothermia. The teacher survived but lost both her feet to frostbite.

Snowstorms can cost money as well as lives. One in February 2011 caused 270 traffic accidents in Illinois. For New York City, it is estimated that each inch of snow costs $1 million to remove, but that is nothing compared with the knock-on effects on trade, commerce, public services and transport. A major snowstorm that hit the city in 1996 was dubbed 'the billion-dollar blizzard'. Even more costly was 'Snowtober', the unseasonally early snowstorm that began on 29 October 2011, which in places dumped 81 cm (32 in.) of snow, lashed by 112 km/h (70 mph) winds, on the northeastern United States. As power lines came down, 3 million people lost electricity. Altogether, 22 people died and the bill for Connecticut alone was estimated at $3 billion. Worse still were the blizzards that hit China in 2008, leaving 4 million people in Chenzhou without power and water for more than a week. There were food and petrol shortages, and nearly 6 million people were left stranded at railway stations. A hotel worker in Chenzhou said: 'It's like we have experienced an air raid or lost a battle.'[7] More than 60 people were killed.

A dust storm blows over Phoenix, Arizona, in 2012.

Blizzard in Manhattan, February 1969.

Then there are ice storms. These happen when a warm front rides over the top of air where the temperature is below freezing and which stubbornly hugs the ground. Raindrops then have to pass through this very cold air, and they freeze as they hit the surface. For the U.S. National Weather Service, an ice storm is one that leads to at least 6 mm (¼ in.) of ice forming on exposed surfaces. One in Texas in January 1940 was said to have delivered ice 15 cm (6 in.) deep. Its weight can do a lot of damage. The ice storm of January 1998 is considered to be one of Canada's worst natural disasters. Over five days, more than 7.5 cm (3 in.) of ice accumulated on surfaces. From eastern Ontario to Nova Scotia, it crushed trees and brought down power lines and pylons, and nearly 16,000 troops had to be dispatched to the stricken areas, the largest peacetime deployment in Canadian history. Federal disasters were also declared in the American states of Maine, New Hampshire, Vermont and New York, and more than 3 million people had their electricity cut off, some for more than a month. Thirty-five people died, and the losses in Canada alone

A 'black blizzard' in South Dakota, one of the many Dust Bowl storms of the 1930s.

were estimated at more than £3 billion; with so many maple trees damaged, production of syrup was cut by 30 per cent.

High winds not only make snowstorms worse but can cause mayhem by hoovering up particles of sand and dust and delivering them as sandstorms and dust storms. A number of famous winds are known for doing this: the Shamal in Iraq and Iran and the Sirocco around the Mediterranean. One of the most redoubtable is the Simoom – Arabic for 'poison wind' – which blows across Arabia and the Sahara and can carry dust storms into southern Europe. Dust particles may be lifted thousands of feet high and carried for hundreds of miles. They get into everything, penetrating buildings, cupboards and machines. In what the National Weather Service called a 'very large and historic' storm in Phoenix, Arizona, in 2011, the moving dust wall reached 3,000 m (10,000 ft) into the air.[8] Drivers were suddenly plunged into pitch darkness as the cloud moved over the city like a giant wave, while in 2013 a dust storm in Nevada caused a 27-vehicle accident.

Plants play an important role in preventing dust storms by holding soil together, but long droughts or excessive grazing or ploughing can erode this protection. The most celebrated dust storms were the ones that afflicted the Great Plains of North America in the 1930s. Native prairie grasses had played the protective binding role when the grasslands were mainly used for grazing, but during the First World War millions of acres were ploughed up to grow wheat. Then came a long dry spell and 'black blizzards' blew away millions of tons of topsoil, burying crops and houses and turning the region into a dust bowl. Eyewitnesses told of walls of dust that looked like great mobile hills, thick enough to block out the sun and cut visibility to a couple of feet. Drifts reached 45 m (150 ft) deep, and thousands of families had to leave their homes – a disaster portrayed in John Steinbeck's novel *The Grapes of Wrath*. Sometimes the soil was carried as far as the East Coast. But these were by no means the only major dust storms in the U.S. In 1977, 800 square miles of the San Joaquin Valley in California were stripped of more than 25 million tons of soil in 24 hours. In Africa the region to the south of the Sahara, the Sahel, had once been grassland, supporting prosperous tribes of herdsmen, but drought, overgrazing and the felling of trees for firewood brought dust storms that left a harsh, desolate landscape stalked by famine.

Sandstorms are more abrasive than dust storms, propelling bigger particles with enough force to scrape glass and scour the paint from cars. The sand does not usually rise more than about

Billboard advertising the film of John Steinbeck's *The Grapes of Wrath* by a migrant camp in California, photographed by Dorothea Lange, 1940.

One of nearly 240 tornadoes that struck the central plains of the United States in late May 2008.

15 m (50 ft) from the ground, but it can bury or even crush buildings – 'a sort of snow that never melts', as one meteorologist put it.[9] One of the deadliest sandstorms of recent years killed a dozen people in Egypt in 1997, but they do not only happen in desert regions. In 2011, a few miles from the Baltic Sea in Germany, a sandstorm caused a motorway pile-up of more than 80 vehicles in which eight people died. Two years later, local authorities in eastern Scotland had to get out snowploughs to clear 1.2-m (4-ft) sand drifts from roads, while an estimated £50,000 worth of crops were destroyed. Sand- and dust storms can also have lethal effects on the environment. In March 2010 a sandstorm that began in Mongolia brought warnings in Beijing that air quality was 'very bad for health' and led to a smog in Hong Kong so severe that schools had to stop children playing outside.[10] And storms can pick up more exotic items than dust and sand. Countries as diverse as the UK, France, Sweden, India, Sudan, the USA and Australia have reported showers of frogs, worms and fish.

This unfortunate habit that storms have of lifting objects into the air reaches its lethal peak with the tornado, which can suck up anything that is not nailed down, and quite a lot of things that are. Tornadoes happen when winds blowing in opposite directions around a strong updraught start a narrow, violent whirl, creating a vortex of very low pressure at the centre. In America the Navajo believed they were ghosts or spirits of the dead, while in Australia Aborigines called the phenomenon a willy-willy, and warned their children there was a spirit inside that would come out if they behaved badly. Some storm systems can spawn many tornadoes. The most prolific recorded came on 3 April 1974, when 148 twisted their way across eleven states of the U.S., killing 329 people and causing damage estimated at $700 million. Tornadoes first appear as little, light grey funnels on the bottom of a cumulonimbus cloud, and if they stay up in the air there is no problem. A few, though, reach the ground. Then their colour turns darker as they pick up soil, branches or bigger objects such as pieces of buildings, cars or animals. These turn the tornado into a kind of giant circular saw as it moves along at a speed of up to 64 km/h (40 mph), cutting through anything in its path. Around the edges of a twister, winds can

The Sahel. Young men carrying straw in Mali.

be spinning at 480 km/h (300 mph). The highest speed ever recorded was 486 km/h (302 mph) at Bridge Creek, Oklahoma, in 1999. The zone of very low pressure at the centre may cause buildings to explode because of the greater air pressure inside them. Fortunately most tornadoes are less than 800 m (½ mile) wide, so it is quite common to see all the houses on one side of a street demolished, while the other side escapes unscathed. They can leave behind odd phenomena, such as pieces of straw embedded in telegraph poles, walls pierced by the branches of trees and splinters of wood driven through sheet metal. Often they last no longer than fifteen minutes before the lethal funnel

Sandstorm at Camp Bastion, Afghanistan, 2013.

dissipates or withdraws into the cloud it came from, though as we will see in Chapter Four, one in 1925 kept going for more than 320 km (200 miles).

Studying tornadoes is very difficult because their winds are so violent that any instruments in their path are liable to be smashed to bits. But because they can pass overhead without touching the ground, a few people have been able to examine one from below at close quarters and live to tell the tale. In 1928 a Kansas farmer named Will Keller described an experience that seemed to last 'a long time', though actually it was just a few seconds. The 'great shaggy end' of the tornado moved directly over him, and everything 'was as still as death'. From the end of the funnel came 'a screaming, hissing sound'. He looked up and saw

> right up into the heart of the tornado. There was a circular opening in the centre of the funnel, about 50 to 100 feet in diameter, and extending straight upward for a distance of at least one half mile . . . Around the lower rim of the great vortex, small tornadoes were constantly forming and breaking away. They looked like tails as they writhed their way around the end of the funnel.[11]

The United States suffers more tornadoes than any other country – about 700 a year. Once again the cause is that collision between cold air from the north with warm from the south. A section of the Great Plains taking in parts of Texas, Oklahoma, Kansas and Nebraska has been dubbed 'Tornado Alley' because it is where twisters are most common, though some of the most savage have hit other states. From 1916 to 1998, tornadoes killed 12,282 people in the U.S., an average of about 150 deaths a year. Since 1998, as the authorities have provided more frequent and more accurate warnings, the total has only twice passed 100, but 2011 was the second worst year on record, with 553 victims. People are killed by flying debris or

Tornado over Lebanon, Kansas, *c.* 1902.

Damage from one of the 74 tornadoes that struck Oklahoma in early May 1999, killing 46 people and damaging or destroying 8,000 homes.

by being dragged across the ground or sucked into the air and then flung to earth. Of 42 people killed at Wichita Falls, Texas, in 1979, sixteen were attempting to flee in their cars only to be blown off the road. The homes of eleven of them escaped any damage. These savage whirlwinds are also seen in many other parts of the world. The UK gets an average of about 60 a year, but fortunately casualties are rare.

Tornadoes at sea are known as waterspouts. They are usually found in tropical and subtropical regions, with more reported in the Florida Keys than anywhere else, though they have also been observed at higher latitudes. One of the biggest ever sighted appeared off the coast of Massachusetts in 1896. Its height was estimated at nearly 1,100 m (3,600 ft), and it was about 255 m (840 ft) wide at its crest and 73 m (240 ft) at its base. It lasted for at least 35 minutes, dispersing and forming three times. A waterspout's swirling winds can reach 305 km/h (190 mph), though most are much slower, and some estimates suggest they can travel at speeds as fast as 80 km/h (50 mph). In 2013 at least

Waterspout in the Mediterranean at Cala Ratjada, Mallorca, Spain.

ten people were drowned when a waterspout sank a boat in the Liguasan River in the Philippines, while in 2011 seven boats were engulfed by the waves as one of these sea twisters ripped through Polruan in Cornwall. Fortunately there were no casualties, and on the whole waterspouts cause far fewer fatalities than tornadoes.

It was in the doldrums, close to the equator, that Coleridge's Ancient Mariner found his vessel becalmed without a breath of wind: 'As idle as a painted ship/ Upon a painted ocean.'[12] And yet these very same hyper-calm regions are the places where the wildest of all storms – tropical cyclones – are born. They can form only over the ocean. Water heats up more slowly than land, but it stays hotter longer, so in late summer and early autumn the ocean's temperature can rise above 27°C (80°F), providing the energy that cyclones need. This means that Atlantic hurricanes normally appear between early June and late November, though a few have been seen in May and December, with the peak season running from mid-August to mid-October. During those months, air disturbances known as African easterly waves cross the Atlantic westwards every few days. They evaporate heat and moisture from the sea, creating thunderstorms. Often these fade away as they encounter cooler water or come ashore, but sometimes they get bigger and join up with others, and the air inside them rises. As it gets hotter, the air pressure falls and winds start to rush faster and faster towards this low-pressure zone. But the rotation of the earth keeps dragging the winds to the right in the northern hemisphere, or the left in the southern, so they can never reach the centre, and the whole system begins to spin – anticlockwise to the north of the equator and clockwise to the south. Once the winds blowing around the centre reach 120 km/h (75 mph), the storm has become what is known as a hurricane in the United States and the West Indies, a cyclone in the Indian Ocean, or a typhoon in China, Japan and the South Pacific.

Viewed from above, the hurricane looks like a huge doughnut of cloud, because at the centre is a relatively quiet, clear area known as the eye. This measures from about 16–95 km (10–60 miles) across and is surrounded by clouds up to 15 km (50,000 ft) high – from the inside, some have described it as resembling a gigantic stadium. The winds that keep spiralling towards the eye run into the thunderstorms of the eye wall, which sweeps them up high towards the stratosphere until the storm can reach a diameter of more than 2,000 km (1,200 miles). Many mysteries remain about why some storms become hurricanes while others peter out, but NASA scientist Dr Owen Kelley sees the eye wall as the most important part of the cyclone – the ingredient that keeps these superstorms going while thunderstorms die away: 'I think of hurricanes as a marathon runner, and I think of thunderstorms as sprinters. They're fast, but then they have to stop', he says.[13]

A tropical cyclone's winds can reach 270 km/h (160 mph), and the whole system moves along at up to 65 km/h (40 mph). Daniel Brown, Senior Hurricane Specialist at the NOAA's

Hurricane Catarina, which made landfall in Brazil in March 2004, photographed from the International Space Station. The eye at the centre is clearly visible.

Atlantic Oceanographic and Meteorological Laboratories, says scientists are now reasonably good at predicting the path that a hurricane will take: 'the tracks of storms are more governed by large scale features in the atmosphere – high pressure or low pressure areas that steer the storms and that is fairly easy to model.' Forecasting their strength, though, is more difficult. In August 1992, for example, Hurricane Andrew turned from a tropical storm into a Category Five hurricane in just over a day, destroying more than 25,000 homes and causing $25 billion of damage in Florida alone. Thirteen years later, Katrina also intensified very quickly. Some scientists say these rapid intensifications may be caused by warm pools that go hundreds of feet

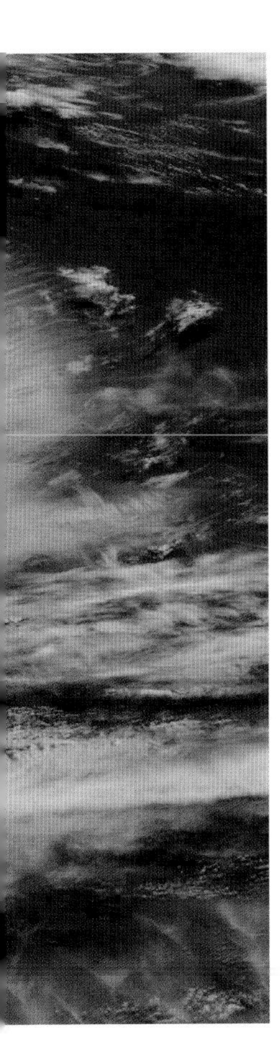

down into the ocean. Others argue that the air in the eye of the storm may sometimes get slightly less humid, enabling it to draw more moisture and heat from the ocean. Dr Kelley suggests sudden intensification can happen when air that has been trapped at the bottom of the eye for a long time, accumulating a lot of moisture and energy, breaks out into the eye wall, igniting 'hot towers' – very powerful updraughts that can soar up to 15 km (50,000 ft) from the earth's surface – though he warns there may well be no single explanation and that 'the answer is probably that it's not the same thing in every case.'[14]

A fully developed cyclone can release energy equivalent to exploding a 10-megaton nuclear bomb every twenty minutes. They do not bring destruction only from the power of their winds. Their low air pressure allows the sea to rise by up to 45 cm (18 in.), adding 7 m (23 ft) or more to the storm surge as it hits land. In 2005 Katrina whipped up waves 17 m (56 ft) high when it crashed into the Louisiana coast, the biggest ever recorded in the region. Then there is the torrential rain that accompanies the storm, falling at perhaps 25 mm every hour. After the cyclone has roared ashore, it is followed by calm and clearer skies, but this is only a temporary respite while the eye passes. Within an hour or two, the wild winds and drenching rains are back. They can sweep across islands and peninsulas or travel along coastlines, but usually when they strike a land mass they soon die out as they lose their source of energy, the warm ocean, and get slowed down by the friction of the land, though in 1938, one managed to penetrate 240 km (150 miles) into New England. From 2005 to 2012 the United States suffered ten hurricanes that caused damage estimated at more than $1 billion, with a total cost of nearly $270 billion, of which Katrina was by far the most devastating at $125 billion. It also cost at least 1,830 lives, but death tolls are often higher in less developed countries, where people live in flimsier homes, and poorer infrastructure makes it more difficult for inhabitants to receive warnings and for rescuers to reach them. So some estimates put the total number killed by a cyclone in Bangladesh in 1970 at 1 million, while up to 140,000 perished when tropical cyclone Nargis hit Myanmar in 2008.

Six weeks after Cyclone Nargis struck in May 2008, debris can still be seen in the streets of Bogale, Myanmar.

A tropical cyclone is one of nature's most terrifying phenomena. A freelance writer named Michielle Beck was caught up in Hurricane Ivan in Florida on 16 September 2004. About midnight, she wrote, 'the wind had begun to literally shriek; it sounded like a woman screaming, and the house creaked and groaned every time a gust hit it.' She believed the building was strong enough to survive, so she moved with her daughter into the master bathroom, the most sheltered place. But it was still terrifying because 'the whole house was shifting with each gust', and 'it sounded like the roof was just going to leave at any minute.' It was twelve hours before Ivan had moved through. The house survived, as did Beck and her daughter, but when she was able to look out, the scene was horrific: 'telephone poles and vending machines lay in the roads, which were littered with debris, roofs were gone, trees (including two in my own back yard) were completely snapped in half.' Less than a mile away, 'houses were washed off their foundations and flooded up to their roof lines.'[15] The following year Alice Jackson, a survivor of Hurricane Katrina, told an even more frightening story. The wind began 'pummelling' her house in the middle of the night. Jackson had been watching

Damage in the Irish Bayou area of New Orleans after Hurricane Katrina in August 2005.

Destruction caused by Hurricane Ivan, Pensacola, Florida, September 2004. More than 90 people died in the United States and the Caribbean.

3099
Briede

with alarm the way a huge pine tree in a neighbour's garden was bending. Suddenly she heard a 'deafening crack' and shouted: 'Run!' She and her family managed to get into the master bedroom in the middle of the house seconds before the tree came crashing through the roof. The walls were heaving, so they rushed around opening windows to relieve the pressure. The house next door had 'turned into what looked like a living, breathing monster. The roof would lift, the house would expand, and then the roof would fall. Finally, the house exploded.' The next day they went for a drive and found power lines lying all over the ground, and 'a big lake where there once had been houses, trees and roads'. After three miles, they were halted by head-high debris.[16]

Katrina, Ivan, Nargis. The practice of giving names to tropical storms goes back hundreds of years. In the West Indies they used to be designated by the saint's day on which they first appeared. Then in nineteenth-century Australia, a meteorologist named Clement Wragge started using the names of politicians he disliked. In the 1920s a system of numbering was tried, but this became confusing when there was more than one storm in an area. Then the authorities tried identifying the tempest by its map reference, but this also proved cumbersome. During the Second World War clarity was vital for military operations, so U.S. meteorologists started to use a woman's name for each storm, sometimes choosing that of a wife or girlfriend. For a few years after the war, unofficial names were given to American hurricanes – one being designated 'Bess' after President Truman's formidable wife. Then in 1953 the U.S. National Weather Service drew up an official slate of female names, arranged alphabetically, to be worked through each season. They had to replace the original choice for 'g' – Gail – with Gilda because of the potential for confusion. The system continued until the late 1970s when women's groups began to protest and the authorities responded by including men's names in the list. Since then hurricanes have been designated by alternating male and female names, though once a name has been applied to a particularly devastating storm, such as Katrina or Ivan, it is retired and replaced. Similar systems now operate across the world.

Designating storms was one thing. What about measuring their intensity? By the early 1700s British sailors were using a rough and ready scale but a century later a Royal Navy admiral, Sir Francis Beaufort, developed something more formal, grading a wind according to the effect it had on a full-rigged man-of-war. There were twelve classifications from zero for 'calm' through ten, with winds of at least 89 km/h (55 mph), for 'storm', to twelve for 'that which no canvas sails could withstand.'[17] The Beaufort Scale was first used officially during Darwin's famous voyage on the *Beagle* in 1831, and from 1838 all ships in the Royal Navy were told to follow it. The U.S. Weather Bureau later added the numbers thirteen to seventeen to cover the more powerful winds experienced across the Atlantic. Then in the second half of the twentieth century, an American structural engineer who was an expert on designing buildings to stand up to hurricanes helped to devise a new method of assessing their potential destructiveness. Herbert Seymour Saffir worked with Robert Simpson, then director of the U.S. National Hurricane Center, to devise the Saffir-Simpson scale. The highest category, Five, is reserved for storms where winds reach speeds of more than 250 km/h (156 mph), and warns that the stricken area will be 'uninhabitable for weeks or months'.[18] (Incidentally, the fiercest wind ever recorded on land was a gust of 407 km/h (253 mph) on Barrow Island off Western Australia on 10 April 1996. The strongest known gust in Britain was 278 km/h (173 mph), recorded at Cairngorm Summit on 20 March 1986.) Similarly, the Fujita Scale, or F-Scale, named after the Japanese-American meteorologist Theodore Fujita, measures the destructive power of tornadoes.

The capacity of storms to do damage is much more obvious than the benefits they bring, but snow, for example, can be very useful. The difference in temperature between the top and bottom of a snow blanket can be as much as 28°C (50°F), shielding seeds and plants against severe cold. And storms play a crucial role in keeping the world's climate temperate by redistributing heat from the equator to the poles. In tropical regions a high proportion of rain comes from storms, and the same applies to some countries outside the tropics, such as Japan. An official survey by the U.S. Department of Agriculture in 1967 concluded that 'the benefits from the rain received from hurricanes far outweigh the damage that may result from the wind or possible local flooding.'[19] In 2012 many U.S. farmers thanked Hurricane Isaac for the 130 mm (5 in.) of rain it brought to relieve a prolonged drought. 'This is exactly the kind of rain we needed', said a farmer in Missouri, who had seen 80 per cent of his corn crop burned off.[20] Storms also help farmers in another way. Lightning's prodigious energy can liberate nitrogen atoms in the air. They then fall to earth with water, and combine with minerals in the soil to create nitrates, which act as natural fertilizers. And fires started by lightning can also be beneficial. They clear undergrowth and debris from woods, turning it into nutrients, and allow sunlight and water to penetrate through to the forest floor, all of which helps seeds to germinate. Hurricane-force winds can do a similar job. In 2011 and 2012 New Jersey's forests took a triple battering from Hurricane Irene, 'Snowtober' and 'Superstorm' Sandy. Big healthy trees were felled by the thousand, but this process, according to some ecologists, brought an explosion of biodiversity in animals, birds, reptiles, insects, plants and fungi. After the Great Storm of 1987 in England, which felled perhaps 15 million trees, the eminent nature writer Richard Mabey said it should be seen 'as an integral part of the workings of the environment, not some alien force'.[21] Twenty years on, the National Trust's head of forestry, Ray Hawes, pointed to what had happened at Toys Hill in Kent. One area of woodland was cleared and replanted after the storm, while another was left as an experimental 'non-intervention' area. The part where nature was allowed to take its course had been more successful, with a far wider variety of trees and flowers. 'At the time', Hawes noted, 'we would have said the wood was destroyed. But we now know that woods are not destroyed, they just change.'[22] Another ecological benefit of storms, noted by Edwin Everham, professor of environmental studies at Florida Gulf Coast University, is that native species tend to survive them better than imported ones. Hurricanes can even save islands. Orrin Pilkey, professor emeritus of geology at the Nicholas School of the Environment and Earth Sciences at Duke University, North Carolina, says they are vital for the survival of barrier islands, making them 'higher and wider' by distributing sand to them. Without this, the islands would get 'skinnier and skinnier' and might disappear altogether.[23]

3 Effects

Storms can divert the course of history, particularly by playing havoc with military operations. One such occasion came in 480 BC, when the Persian emperor Xerxes the Great decided to conquer Greece. The ancient Greek historian Herodotus, the 'father of history', tells how Xerxes assembled an army the like of which had never been seen before, drawn from all over Asia. They were led by a warrior elite, the Ten Thousand Immortals, clad in gold.[1] Some modern estimates put the army's size at 360,000. They advanced to the ancient city of Abydos on the Dardanelles, the straits that divide Asia from Europe, where, says Herodotus, 'there is a rocky tongue of land which runs out for some distance into the sea'. That still left a gap the best part of a mile wide, so the leaders of the expedition decided to build two pontoon bridges. But as soon as they were finished, 'a great storm arising broke the whole work to pieces', and the expedition was halted in its tracks. When Xerxes heard what had happened, he flew into a rage, and ordered that the Hellespont be given 300 lashes along with a personal message from him: 'Thou bitter water, thy lord lays on thee this punishment because thou hast wronged him without a cause, having suffered no evil at his hands. Verily King Xerxes will cross thee, whether thou wilt or no.'[2] The unfortunate folk in charge of constructing the first bridges were beheaded, then a new one was built, and this time the great army passed over successfully. The Greece Xerxes sought to conquer was actually a collection of many small city-states, and he hoped to profit from their disunity, but,

Otto Albert Koch, *Battle of the Teutoburg Forest*, 1909, oil on canvas.

numbered only about 40,000. But once again, a fierce typhoon struck. All but a few hundred ships were sunk and at least half of the Mongol warriors were drowned. Most of the survivors were hunted down and killed by the Japanese, so that fewer than one in five escaped. In Japan it was said that 'a green dragon had raised its head from the waves' and the destruction was so great that 'a person could walk across from one point of land to another on a mass of wreckage.'[16] So the country never was conquered by the Mongols, and the storms that had saved it were dubbed 'divine winds' or kamikaze – a name used in the Second World War for pilots who mounted suicide attacks on enemy ships.

The biggest and bloodiest battle on British soil since Roman times took place on Palm Sunday, 29 March 1461, during the Wars of the Roses, and once again a storm played a crucial role. The Yorkist king Edward IV had just grabbed the throne, deposing his Lancastrian rival, Henry VI, whose forces had withdrawn to the north. A week after being proclaimed king, Edward set off in pursuit, and caught up with the enemy near the village of Towton in Yorkshire. Each army is said to have comprised at least 30,000 men. They took up positions on a plateau about 50 m higher than the surrounding area. In most parts the land fell away gently, but there was quite a steep slope down to a small river called the Cock Beck, which was, according to the Tudor chronicler Matthew Hall, 'not very broad, but of a great deepness'. Hall records that as the two armies first sighted each other, sleet and snow began to fall. The devout but ineffectual Henry VI had wanted any hostilities postponed because of the holiness of the day, but no one else was much bothered. Edward, in particular, was keen to get on with it, as he had a strong wind at his back, and soon the Lancastrians could scarcely see because its 'violence' was driving the snow into their faces. The wily Lord Fauconberg, who commanded Edward's vanguard, got his archers to fire a volley and then retreat a few paces. Stung by the Yorkist arrows, the Lancastrians replied with a series of salvoes, but, firing into the wind, they were wasting their time: 'all their shot was lost and their labour vain', as their arrows fell harmlessly short. Nor could they see that, as their fire slackened, Fauconberg sent his

Utagawa Kuniyoshi (1797–1861), *Nichiren's Destruction of the Mongol Fleet, 1281*, colour woodblock print. For the second time, Kublai Khan's invasion of Japan is thwarted by a typhoon, which the Japanese dub a kamikaze, or 'divine wind'.

men forward to pick up the Lancastrian arrows and return some to the enemy with deadly effect, while from others he built an improvised barricade against any Lancastrian advance.[17]

Constantly losing men to the arrows carried on the Yorkist wind, the Lancastrians had little alternative but to attack. Edward IV had given his troops orders that no quarter was to be asked or given. A 'deadly battle and bloody conflict' now followed in the blizzard, lasting for ten hours. In addition to those who fell by the sword, many men collapsed from exhaustion and were trampled to death. When Yorkist reinforcements arrived, the Lancastrians finally broke. 'Discomfited and overcome, and like men amazed', they fled – most of them heading for Tadcaster, but in the way at the bottom of that steep slope was the Cock Beck, swollen by snow and sleet. This now became a bloody bottleneck. Many fleeing Lancastrians were cut down by pursuing Yorkist cavalry, but others were 'drenched and drowned' in the beck, until 'men alive passed the river upon dead carcasses.' The battle had ended up as a murderous rout, and 'the chase continued all night, and the most part of the next day.' The lowest

contemporary estimate for the number of casualties was 28,000 – 20,000 of them Lancastrians.[18]

Britain's biggest battle since Roman times, at Towton in Yorkshire, was fought during a blizzard in 1461. This engraving was published *c.* 1878.

Britain's bloodiest battle it might have been, but Towton is not a household name, unlike the Spanish Armada. And the part played by storms in its defeat is well known. One of the medals Elizabeth I had minted to celebrate the victory carried the legend 'God breathed and they were scattered.'[19] This fitted perfectly with the propaganda message that the Lord had thrown in his lot with plucky English Protestants against the overweening Catholics of the world's most powerful empire. If God really was intervening, he got to work early. Right from the start, storms meddled with the Spanish king Philip II's plans. The great armada of 130 ships set out from Lisbon, then ruled by Spain, on 9 May 1588, but immediately ran into gales and had to anchor for three weeks, with pilots complaining to the admiral, the Duke of Medina-Sidonia, that the weather was more like December than May. Even when the fleet got going on 30 May, it took thirteen days to cover the first 300 km (190 miles) in what was supposed to be the best month of the year for sailing.

Then on 19 June came a fearful tempest that scattered the ships off La Coruna. Things were so bad that the Duke wrote to the king suggesting he postpone the invasion until the following year, but Philip ignored his plea. Some of the Spanish ships had been driven as far as the Isles of Scilly, and by the time the admiral had gathered his fleet together, done the necessary repairs and was ready to depart again, it was 21 July.

So their enemy was well prepared when at last, on 29 July, the Spanish Armada was sighted off Cornwall. Its first task was to go to the Netherlands to rendezvous with 30,000 redoubtable Spanish troops – the men who were to conquer England. Getting them into their transports was expected to take six days, and while they were still embarking, the English won a decisive victory off Gravelines, sinking or driving aground three ships of the armada and severely damaging many others. Unfavourable winds and the menacing presence of the English fleet, which had faster vessels and better guns, then made it impossible for Medina-Sidonia to meet with the soldiers' transports and escort them across the Channel, destroying the invasion plan. The admiral decided the best service he could now perform for his king was to get as many ships home safely as he could.

That meant a perilous journey around the coasts of Scotland and Ireland, being buffeted by storms and gales. These tempests turned a defeat into a disaster. On 17 August the *Gran Griffon* was driven onto Fair Isle, where the crew were stranded for the whole of what must have been a very cold and unpleasant winter. On 20 September three ships anchored in Sligo Bay when a fierce storm blew up. One of the captains told how they were driven ashore: 'Such a thing was never seen; for within the space of an hour all three ships were broken to pieces, so that there did not escape 300 men, and more than 1,000 were drowned.'[20] Altogether, at least seventeen vessels came to grief off the coast of Ireland. A galleass, the *Girona*, was packed with 1,300 men who had crowded aboard when their own ships were lost. Then she had her rudder smashed by a gale that blew her on to the Antrim coast close to the Giant's Causeway, where she split apart on a reef in pitch darkness. Only nine men survived. Of the 130 ships that had set out from Lisbon,

Medina-Sidonia was able to get only perhaps 85 safely back to Spain, and up to 15,000 of the 27,000 men who had embarked in May never saw their homes again. The fate of some of the lost ships remains unknown to this day.

By the time the Second World War came along, a lot more was known about predicting the weather, and by the war's end the RAF had six squadrons devoted to gathering meteorological data, with the USA and Germany being equally assiduous. Hidden in a haystack on the Berkshire Downs was the British Meteorological Service's top secret Thunderstorm Location Unit, which was able to record all major flashes of lightning up to 3,200 km (2,000 miles) away. Not only could lightning itself be lethal to aircraft but its presence might indicate treacherous air currents or potentially fatal hazards like severe icing, or the likelihood that bombing targets would be obscured by thick cloud. The information gathered saved the lives of many pilots, but much

Defeat of the Spanish Armada, 8 August 1588, by Philip James de Loutherbourg, 1796, oil on canvas. The picture shows the Battle of Gravelines.

A U.S. bombing raid on Germany, 9 October 1943.

of it was kept secret even from the British people, whose weather forecasts were not allowed to mention things like gales or snow. The Thunderstorm Location Unit was hooked up to the Central Forecasting Station at Dunstable. The Germans were always trying to find it but never did. It was there that the crucial weather forecast for D-Day was issued, resulting in Eisenhower postponing the invasion from 5 June to 6 June to take advantage of a brief lull in the stormy weather that was threatening to make the assault impossible.

Greater knowledge of the weather, though, did not make forecasting infallible and Allied fleets in the Pacific were caught in typhoons off the Philippines in December 1944 and off Okinawa in June 1945. This maddening refusal of storms to behave as expected meant that a crucial American mission ended in disaster more than three decades later. In April 1980 the Iranians had been holding 53 Americans hostage for more than five months, and diplomatic efforts to free them had run into a blind alley, so President Jimmy Carter decided to launch an extremely risky rescue attempt. Eight helicopters were to snatch the hostages from Tehran, a city surrounded by hundreds of miles of desert and mountains, and a long way from any country friendly to the U.S. If the mission was to be successful, it would need virtually clear weather, and that was exactly what its commander, Lieutenant-General James Vaught, was promised on 23 April. Instead the helicopters ran into a haboob, a dust storm. Two were knocked out, reducing the force to six, the bare minimum needed to pull off the rescue. Then another lost part of its hydraulic system, and was declared too dangerous to use. General Vaught recommended abandoning the operation, and the President agreed, but as they were leaving one of the helicopters crashed into a tanker aircraft parked on the ground. It burst into flames, killing eight Americans. The wreckage was abandoned, allowing top-secret documents to fall into the hands of the Iranians. Carter went on television to accept the blame for what was starting to look like a fiasco. The hostages were not released until January 1981, by which time the president had suffered a crushing election defeat at the hands of Ronald Reagan.

But it is not only by confounding the plans for military operations that storms can change the course of history. During the 1780s France suffered a series of poor harvests. In 1787 there were floods, then in the spring of 1788 many places suffered drought. The last straw was added to the people's backs on 13 July of that year, when a devastating hailstorm swept across the north of the country. Men and animals were killed, there were reports of trees being torn up by the roots and many of the crops ripening in the fields were destroyed. Grain and bread really were the staff of life in those days, with the poor spending up to half their income on these two commodities. In Nantes the cost of a loaf went up to five sous, more than four times what it had been in the early 1760s, while in Paris prices reached famine level. Popular discontent, rife enough by this time, grew even more strident, but that was not the only problem. The ruined harvest meant a sharp dip in tax revenues at a time when the government was already facing a financial meltdown, and on 16 August it went bankrupt. The crisis meant that Louis XVI had to call the nearest thing France had to a parliament, the Estates-General, for the

An aerial view of the islands of Bermuda, which may have inspired Shakespeare's *The Tempest*.

first time in 175 years, setting the country on the road to the revolution that would cost him his throne and his head.

A more bizarre instance of storms tweaking history came in the early years of the seventeenth century when they helped create what is today a holiday paradise. In 1609 the settlers at Jamestown, Virginia, the first permanent English settlement in North America, were in serious trouble. What with quarrels among themselves, fights with native peoples and disease in the swampy, unhealthy country, their original number of 143 had declined to just 38. So the Virginia Company dispatched a convoy of nine ships, led by the *Sea Venture*, carrying 150 would-be colonists – men, women and children. One of the *Sea Venture*'s passengers was a writer named William Strachey. His account of the voyage records that on 24 July, when they were 965 km (600 miles) off the American coast, they ran into a tempest which split them from the rest of the ships: 'For four and twenty hours the storm in a restless tumult, had blown so exceedingly, as we

4 Events

What was probably the deadliest storm in history was not one of the strongest. With wind speeds peaking at 185 km/h (115 mph), it was classified as Category Three on the Saffir-Simpson hurricane scale, on which the fiercest are Category Five. But the place it struck on 12 November 1970, then East Pakistan, constituted a disaster waiting to happen: more than 100 million people, mainly poor, crowded onto land mostly less than 3 m (10 ft) above sea level in the Bay of Bengal, which was devastated by cyclones at regular intervals. In 1737 more than 300,000 were said to have been drowned, in 1963 about 20,000 and in 1965, 40,000. But what became known as the Bhola Cyclone was the deadliest of all.

A storm warning had been issued on the radio, but the wording was vague, and three weeks earlier another alert had been followed by a cyclone that fizzled out before it hit land. So most people were asleep when, around midnight, the tempest sent a great wave crashing over islands off the coast. According to a farmer on one of the bigger ones, Manpura, the first sign was a 'great roar'.[1] Then, in the pitch darkness, he saw a glow. As it came nearer and nearer, he realized it was the crest of a 6-m (20-ft) wave. He and his family had a house more solid than most, and were able to take refuge on the roof. For five hours, they endured the lashing wind and rain, while all but four of their fellow islanders' 4,500 bamboo and thatch huts were destroyed, and 25,000 out of Manpura's 30,000 inhabitants perished. On thirteen small islands, it was reported that no

A woman walks through demolished buildings in the aftermath of the tornado that struck Brahmanbaria, Bangladesh, in March 2013, killing more than 30 people.

one was left alive. On the biggest island, Bhola, one in five of the population of 1 million was drowned; the heaviest death toll was among children, who were too weak to hang on to trees. On Shakuchia a 40-year-old rice farmer clung to a palm tree with his wife as, one by one, their six children were torn from their grasp. Then the farmer too was taken by the waters. In her despair his wife let go, but he managed to grab her and take hold of another tree, which they held on to until dawn when the water began to recede. By then there were bodies everywhere, lying on the ground or hanging from trees. On Jabbar Island one old man had to pile the remains of 52 of his relatives into a mass grave. There were so many bodies that some were put on rafts and floated out to sea, but often they were washed back again. Even a week after the storm, rescuers could not move without stepping on corpses. One reporter in the Ganges delta saw 'at least 3,000 bodies littered along the road. Survivors wandered like mad people, crying out the names of their dead ones.'[2]

There were some astonishing escapes. Three days after the storm, a wooden chest was washed in from the sea. Inside were six children aged twelve and under, still alive, and their grandfather, who had died of exposure. He had put them in the chest before climbing in himself. Altogether a million cattle were killed and a million acres of paddy fields were swamped, leaving desperate villagers searching the mud for single grains of rice. Springs were poisoned by seawater and rotting corpses, and there were fears of typhoid and cholera. Within a few days the international relief effort got into its swing, ferrying supplies to the stricken region, though not enough to prevent many more deaths from disease, exposure and starvation. The government in West Pakistan, more than a thousand miles away, was accused of being grudging and dilatory in its response. The death toll is generally regarded as being between 300,000 and 500,000, though some believe the true total to be more than 1 million, and millions more were left homeless. There had been plenty of discontent before the cyclone, but after it the mood hardened into a full-scale campaign for independence. There followed a bloody civil war, in which perhaps another 3 million people were killed, before the new nation of Bangladesh emerged.

The Bhola Cyclone also has a place in cultural history: it inspired the first great international fundraising rock show. The Concert for Bangladesh of 1971 featured stars including George Harrison, Ringo Starr, Bob Dylan, Eric Clapton and Ravi Shankar, and included a song specially written by Harrison about the country's plight. It was the forerunner of events such as Live Aid. After independence, Bangladesh embarked on a programme of building storm shelters, but it could do nothing about the country's geography, and in 1991 another cyclone, this one Category Five, killed 138,000 people.

Bangladesh's neighbour Myanmar suffered a similar number of casualties when it was hit by Cyclone Nargis in 2008. The storm struck the densely populated rice-growing region of the Irrawaddy delta on the night of 2 May. India's meteorological agency had passed on information about how devastating Nargis was going to be, and Myanmar's military junta did broadcast some warnings, but they failed to organize an evacuation or take other steps to limit casualties. Perhaps they were preoccupied with preparations for a referendum on a new constitution, due to happen on 10 May. The Category Four cyclone began by hurling a 3.5-m (12-ft) storm surge on the low-lying villages of the delta, sweeping away whole coastal communities. In the village of Pyin Ma Gone, a 52-year-old farmer said: 'The wind became so strong it ripped my house apart. My wife was blown away to the other side of the river. We never saw her again.'[3] Ten members of his family died. Sean Keogh, a British doctor working for a medical aid agency, described the devastation in the delta: 'Some families have been so completely wiped out, there is no-one to bury the dead. They are hanging from trees and trapped on posts.'[4] In the town of Bogale 95 per cent of the houses were said to have disappeared, with 10,000 people killed. In Kyaiklat, which had a population of about 50,000, most homes were damaged, but one shop owner said he survived by taking refuge in a concrete house, while some of his neighbours ran to the monastery.

As in Bangladesh, the government's relief efforts were heavily criticized. At Kungyangon local people said it was not until four days after the storm that a group of soldiers arrived and handed out some meagre rations, then lounged about by the roadside. There were complaints that whether supplies were delivered or not depended on the whim of the army officer in charge of the district. As hunger followed in the wake of the cyclone, children, especially orphans, were at particular risk. The amount of aid arriving, said Dr Keogh, was 'not enough and it's not quick enough.' The natural hazards of continuing torrential rain combined with the primitive infrastructure would have been handicap enough for the rescue operation, but in addition the paranoid junta, which had bloodily suppressed protests earlier in the year, was reluctant to allow foreign ships to land or foreign aircraft to fly in supplies, and slow in granting visas to aid workers. It was not even very keen on allowing its own people to go to help, and one of the country's leading comedians was arrested for his pains. The upshot of this approach was that the biggest emergency relief supplier, the UN World Food Programme, whose helicopters were not allowed to fly in until a month after the disaster, said it was only able to deliver a fifth of the supplies that were needed. The U.S. Secretary for Defense, Robert Gates, accused the regime of 'criminal neglect'.[5] One thing the generals did assiduously, though, was to cover up labels showing that aid had come from abroad.

The junta also tried to stop foreign journalists getting to the disaster zone, but one BBC reporter managed to reach what had been the fishing village of Uomiou two weeks after Nargis struck. He found twenty survivors in the only house still standing. Even that had had one of its walls knocked down. One of the survivors was a 77-year-old woman who had lost her children and her grandchildren, and reported tearfully that the few people left were starving. In the town of Tabitha a young girl was turned away from a refugee camp because, officials told her, she 'was not on the list of survivors'.[6] A month after the cyclone 2.4 million people were said to be still homeless. For many, Buddhist monks were the only source of help, and people would carry sick relatives on their backs for miles through mud and rain, or paddle for hours on stormy rivers to reach them. Monks had died in the storm, too, along with villagers and now comforted survivors in the muddy squalor it had left. One woman, who had lost everything, said she was thinking about killing herself until she heard that a monk had opened a clinic 10 km (6 miles) upriver. 'In my entire life, I have never seen a hospital', she said. 'So I came to the monk. I don't know where the government office is.'[7] Official television pictures portrayed a fantasy world in which adequate supplies were distributed in an orderly way, while unauthorized video shot by local people showed survivors packed into monasteries sitting shoulder to shoulder on the floor waiting for hand-outs of food and water.

Whatever the junta's shortcomings in dealing with the cyclone, it successfully held its referendum on the due date, though it allowed voting to be delayed for two weeks in the stricken areas. The new constitution secured 93 per cent approval, but within four years the generals had released their iconic arch-opponent, Aung San Suu Kyi, and held free parliamentary elections. Some believe the storm played a crucial role. Andreas List, head of the EU Office in Yangon, said: 'Cyclone Nargis was a turning point. After it, the government realised they needed international help.'[8]

The deadliest ever Atlantic hurricane made landfall in Barbados on 10 October 1780. According to a contemporary account in *The Gentleman's Magazine and Historical Chronicle*, the previous evening had been 'remarkably calm; but the sky surprisingly red and fiery'. Then during the night, 'much rain fell'. By morning, the wind had 'increased very much'. By four o' clock in the afternoon, about 25 ships had been driven out to sea, and by six in the evening, what became known as the Great Hurricane 'had torn up and blown down many trees'. At Government House they took 'every precaution', barricading doors and windows, but 'it availed little'. The governor, Major-General John Cunninghame, and his family took shelter in the centre of the building, where the walls were a metre thick and of 'prodigious strength', but even there they did not feel safe, with the wind becoming so fearsome it almost took off the roof. Before midnight they had retreated to the cellar, but torrential rain started to flood it, and they ran outside, believing this to be safer than staying in the building. The armoury had by now been 'levelled to the ground'. Taking shelter as best they could, they waited for daybreak, 'flattering themselves that with the light they would see a cessation of the storm', but no such luck. The tempest seemed as fierce as ever, and nothing could be 'compared with the terrible devastation that presented itself on all sides: not a building standing. The trees, if not torn up by the roots, were deprived of their leaves and branches.' What had been a luxuriant, fertile island was transformed 'in this one night, to the dreariest winter'. Many people, 'whites and blacks together', had been buried beneath rubble, with others washed out to sea. Most buildings were destroyed or seriously damaged, and a nineteen-pounder gun had been carried 130 m by the wind.[9]

While the storm was raging, the commander of the Royal Navy's Leeward Islands fleet, Admiral Sir George Rodney, had been away fighting the American rebels. When he arrived in Barbados a few weeks later, he was astonished at

> the dreadful situation of this island and the destructive effects of the hurricane. The strongest buildings and the whole of the houses, most of which were of stone, and remarkable for their solidity, gave way to the fury of the wind and were torn up to their foundations.[10]

He added: 'Had I not been an eyewitness, nothing could have induced me to have believed it.' Meteorologists have since estimated that this was a Category Five hurricane with wind speeds of more than 320 km/h (200 mph). The number killed on the island was put at 4,500. Major-General Cunninghame wrote: 'The loss to this country is immense: many years will be required to retrieve it.' After Barbados, the hurricane moved on to St Lucia where all but two of the houses in the port city of Castries were blown down. Five of Admiral Rodney's ships were destroyed and nine others severely damaged. One was driven against the naval hospital, killing everyone on board and everybody in the building. The island's death toll was estimated at 6,000, while on St Vincent more than 580 out of 600 houses at Kingstown were destroyed. At Grenada nineteen Dutch ships were wrecked, and a French fleet of 40, sent to help the Americans in the War of Independence, was caught off Martinique, where about 4,000 sailors were drowned. All the houses in the town of St Pierre were destroyed, and perhaps 9,000 people died. A storm surge on the Dutch island of St Eustatius killed up to 5,000 more, and the storm also hit Dominica, Guadeloupe, St Kitts, Puerto Rico, the Dominican Republic and Bermuda, killing up to 30,000 people altogether. A leading American rebel, James Duane, described the hurricane as 'the worst disaster since the Deluge', speculating that it might have struck a fatal blow against the Royal Navy in the War of Independence.[11] Within eighteen months the colonial power did indeed decide to abandon the struggle, but the havoc caused by the hurricane was a contributory rather than a decisive factor.

Not only did Bangladesh suffer the deadliest storm of all time, it was also the victim of the most lethal tornado. What became known as the Saturia–Manikganj Sadar tornado struck these two districts around 65 km (40 miles) northwest of the capital, Dhaka, at about half past six on the evening of 26 April 1989. In Hargoz village every house was destroyed. Local people described huge trees being sent flying through the air like kites, and cattle being snatched up and hurled down hundreds of metres away. 'It looked as though the whole village was being uprooted', said a 45-year-old woman.[12] Within minutes there were 'hundreds of bodies lying around and thousands of people crying for help'. According to one local council member, 'People were blown far away and some bodies were found one or two miles from the village.' Another added that Hargoz had been 'turned into a mass grave', while doctors at Manikganj hospital said they had treated more than 1,000 people who had lost limbs or suffered broken bones. Over an area of more than 50 square km, virtually every home was flattened.

Rescuers were hampered by heavy rain, and in the days that followed the tornado there was much criticism of the slowness and inadequacy of the relief effort. A businessman in Hargoz

John Thomas Serres, *The Wreck of the HMS 'Deal Castle'*, 1780, oil on canvas. One of the many ships to come to grief during the Great Hurricane of 1780, the deadliest ever seen in the Atlantic.

complained that it amounted to nothing more than 'a drop in the sea'. Nearly a week after the tornado struck, a journalist reported the village was 'almost deserted, the survivors having fled in search of shelter and food. Toys and household goods lie scattered amid heaps of debris, reminders that this was once a thriving community.' Beside the graveyard, a 65-year-old man wept as he waved a bamboo pole to try to scare off vultures. Many survivors in the area were hungry and an official from the Bangladesh Red Crescent Society said others were ill after eating rotten food or drinking contaminated water. More than 100 were injured in a stampede in a remote area of Manikganj district when desperate people besieged a relief lorry. Overall the tornado cost the lives of about 1,300 people and left another 50,000 homeless.

Four of the world's five deadliest tornadoes happened in Bangladesh. The only one to afflict another country, and the third deadliest of all, was the Great Tri-State of 18 March 1925, which pursued a trail of devastation up to a mile wide across the American states of Missouri, Illinois and Indiana. More than 25 towns, including Murphysboro, Illinois, where 234 died, found themselves in harm's way. At De Soto, Illinois, 100 people were killed and 300 injured out of a total population of just over 700. One of the most vivid descriptions came from a woman who had been sitting in a restaurant at Gorham, Illinois. As it started to rain, she decided to go home, but when she opened the door, she saw 'a great wall that seemed to be smoke, driving in front of it white billows that looked like steam. There was a deep roar, like a train, but many, many times louder. The air was full of everything – boards, branches of trees, pans, stoves, all churning around together. I saw whole sides of houses rolling along.' She was flung back into the restaurant as 'the building rocked back and forth and then it began to fall in.' Something knocked her unconscious, and when she came round she was buried under debris, but close by was 'the body of a red cow which seemed to be holding some of the weight off me'. She was pulled free by a man looking for his sister. One of her friends lay dead with a wound to her head. As soon as she could, she raced to the school and found her children – injured but alive. A great crowd of people were trying to dig out their family members: 'Children were screaming and crying. Mothers and fathers were weeping silently.'[13] The Great Tri-State Tornado lasted for three and a half hours – a record, as was the 352-km (219-mile) distance it covered

The ruins of Longfellow School in Murphysboro, Illinois, where 17 students were killed in the Tri-State Tornado of 18 March 1925, the most deadly in U.S. history.

at unusually high speed. It killed 695 people, more than 600 of them in Illinois.

Probably the deadliest blizzard in history struck the Ardekan area of southern Iran during the first week of February 1972. The villages of Kakkan and Kumar were both buried under snow-drifts up to 8 m (26 ft) deep. Further north in the hamlet of Sheklab on the Turkish border, rescue workers used a respite in the storm to try to reach villagers entombed in their homes. It is said they burrowed for two days but all they discovered were eighteen corpses and no survivors. At the Koheen Pass, 280 km (175 miles) northwest of Tehran, five people were found dead after being trapped in a car for five days as temperatures fell to –25°C. In some places, army helicopters dropped bread and dates on the snow hoping that survivors might be able to tunnel to the surface and find them. The final death toll was estimated at 4,000. The unparalleled severity of the Iran blizzard is underlined by the fact that the next deadliest on record was the one that struck Afghanistan in February 2008. The number of deaths was 926, with more than 300,000 cattle also killed, while more than 100 people had to have limbs amputated because of frostbite – mostly shepherds who had got lost in the mountains.

The deadliest documented hailstorm felled 246 people in just two minutes on 30 April 1888 at Moradabad in northern India. Some of the victims were pounded to death, but most suffocated or were frozen as they lay buried under drifts, said to be 60 cm (2 ft) deep in some places. The *Times of India* reported: 'The shape of the stones generally was a flat oval, very few being round like ordinary hail.' It added that, 'incredible as it may seem', one stone picked up in the garden of the hospital weighed 700 g (1 ½ lb), while, 'more extraordinary still', another found near the telegraph office, turned the scale at 900 g (2 lb). The newspaper assured readers they had been weighed by 'two gentlemen of unquestionable veracity'.[14] In addition to the humans killed, more than 160 cattle, sheep and goats perished, while most houses had their roofs destroyed, trees were uprooted and at Government House 200 windows were broken. The *Times* of London wrote that 'masses of

Niya, the 'Pompeii of the Silk Road', said to have been buried by a sandstorm *c.* AD 400.

frozen hail remained lying about long after the cessation of the storm.'[15]

But more recently, Moradabad's dubious distinction of being the scene of the world's deadliest hailstorm has been called into question. By the remote lake of Roopkund, 4.8 km (16,000 ft) up in the Himalayas in 1942, a forest ranger found almost 600 human skeletons. They were all of people who had been killed by blows to the head. The place was quickly dubbed 'Skeleton Lake' and the first theories were that these were victims of a landslide, or even a mass suicide. Because the local climate is so cold, an expedition in 2004 was able to find some hair and flesh as well as preserved clothing. This enabled them to be confident that the skeletons dated back to around AD 850. They also seemed to fall into two distinct groups: one was made up of taller folk who were closely related, while the others, who were shorter, appeared to be local people. The expedition concluded that the taller ones were pilgrims who had hired locals as guides and porters. They also made a more startling discovery. The cracks in the skulls and the shoulder injuries of the victims appeared to have been inflicted from directly above by

from the church. The explosion destroyed a sixth of the city, as well as 'greatly injuring' the remainder and killing, it is claimed, 3,000 people. Similar accidents happened at Malaga in 1780, on Sumatra in 1782, in Tangiers in 1785 and in Luxembourg in 1807, but the deadliest ever explosion sparked by lightning was probably the one at the church of St John on the island of Rhodes on 6 November 1856. This ancient cathedral of the knights of Rhodes had a 'large collection' of gunpowder in its vaults, and the explosion destroyed a 'large portion' of the town, causing a death toll estimated at 4,000.[18] The *Times* sadly recorded: 'The fine old castle and cathedral of the Knights of St John are now a mass of hideous ruins.'[19]

Poster for *Miracles Still Happen* (1974), one of the films telling the story of Juliane Koepcke, the sole survivor of the crash in Peru of LANSA Flight 508 on 24 December 1971.

Lightning can also bring down aeroplanes, and the deadliest such accident happened on Christmas Eve 1971. Sometimes, in air crashes, there is no one left to tell the tale, but on this occasion there was one survivor from the 92 people on board. Juliane Koepcke from Germany, aged seventeen, and her mother had boarded a Lockheed Electra turboprop, operated by the Peruvian airline LANSA, in Lima for a flight to Pucallpa in the Amazonian rainforest. They were on their way to join her father for Christmas; her parents, both well-known zoologists, ran a research station in the jungle. About 25 minutes into the flight, the aircraft entered very heavy, dark cloud and began to shake. Juliane said that at first she was not afraid. Then suddenly there was a bright flash and the Lockheed went into a nose dive. She and her mother held hands while people screamed and 'Christmas presents were flying around the cabin'.[20] Then they saw a 'very bright light' on one of the engines.[21] Juliane's mother said calmly: 'That is the end, it's all over.' That was the last thing her daughter heard from her. Then everything went pitch black, and Juliane found herself outside in the air. 'Suddenly', she said, 'there was this amazing silence. The plane was gone.'[22] An accident investigation later found that one of the fuel tanks on the Lockheed had been hit by a bolt of lightning, and that this had torn off a wing. Juliane plummeted for 3 km (almost 2 miles), strapped to a row of seats: 'I was flying, spinning through the air and I could see the forest spinning beneath me.' She crashed through the thick jungle canopy, which broke her fall. When she awoke the next morning, her first thought was: 'I survived an air crash.'[23] Juliane had broken her collarbone and suffered deep cuts to her legs. She shouted for her mother, but soon realized she was alone. As rescue crews searched in vain for the missing aircraft, remembering advice from her father to follow a waterway if ever she was lost, Juliane began to wade down a creek. She passed another row of seats with the bodies of three female passengers still strapped in and rammed head first into the ground. The only food she had was a bag of sweets. The creek eventually led her to a river, and after ten days she came to a hut, where she stayed. The next day a group of Peruvian lumberjacks found her, treated her wounds and got her to hospital. The dead body of her mother was discovered two weeks later. The famous German director Werner Herzog would later make a documentary film about Juliane's escape titled *Wings of Hope* (2000).

Peter Monamy, *The Opening of the First Eddystone Lighthouse in 1698*, *c.* 1703, oil on canvas.

The deadliest storm ever to hit Britain is still known simply as the Great Storm more than 300 years later. The tempest of October 1987 was given the same title, but mercifully it killed just eighteen people; its predecessor from 1703 cost the lives of perhaps 8,000. One of the first to experience it was the vicar of St Keverne on the Lizard peninsula in Cornwall at about eleven o' clock on the night of 26 November. He said the wind was blowing so fiercely that 'the country hereabout thought the great Day of Judgment was coming.'[24] One of the storm's first victims was the theatre owner, engraver and engineer Henry Winstanley, who had built a lighthouse on the treacherous Eddystone Rocks 23 km (14 miles) off Plymouth. It had a stone base, while the upper part was made from wood bound with iron straps. Winstanley considered his lighthouse indestructible and once expressed the hope that he might be in it during 'the greatest storm that ever was'.[25] On 26 November 1703 his wish was granted. He and two colleagues were inside the structure doing repairs when 18-m-high (60-ft) waves began dashing against it. All we know of his fate is that by the time the Great Storm had abated no trace could be found of him, his companions or his lighthouse. Another eminent victim was the Bishop of Bath and Wells. He and his wife both perished when a chimney stack fell through the roof as they lay in bed in the bishop's palace at Wells.

There was damage all along the south coast. Brighton was inundated, and at Shoreham the ancient Market House was 'blown flat to the ground and all the town shattered'.[26] Other coastal towns, such as Portsmouth, 'looked as if the enemy had sacked them and were most miserably torn to pieces'.[27] A huge tidal wave was said to have raced up the Severn, knocking birds out of the air and hurling them against buildings, while much of Bristol was flooded, so that valuable goods stored in warehouse cellars were ruined. A church and other buildings were also blown down, killing a number of people. In Wales, Cardiff had a big breach gouged in its city wall, while in Swansea most houses lost their roofs. Churches were damaged as far east as Maldon in Essex, with many spires toppled, including Kent's tallest at Brenchley. The county also saw more than 1,000 houses and barns completely flattened. Near Oxford a clergyman saw a tornado 'marching directly with the wind'. It looked like an elephant's trunk, 'only much bigger. It was extended to a great length, and swept the ground as it went, leaving a mark behind.' Close to Hinksey, it knocked a man over, and encountering an oak tree, 'snapped the body of it asunder'.[28]

The storm smashed windows and blew down pinnacles at King's College Chapel in Cambridge, while at Northampton, sheets of lead on church roofs were 'rolled up like a scroll'. Three windmills were destroyed there. At one, 'the mighty upright post below the floor' was 'snapped in two like a reed'. There were some remarkable escapes. At Highbridge on the Somerset Levels, rising waters caused a house to collapse, killing a man and a woman who lived there, but their child floated out of the building in its cradle and was later found alive. At Charlwood in Surrey a miller was woken by the storm and rushed out to his mill to try and save it. When he got there, he realized he had left his key at home, and had to go back. His forgetfulness saved his life, because when he reached the mill a second time, he found the storm had blown it away. Livestock were killed in huge numbers, particularly around the Bristol Channel and the Severn, where perhaps 15,000 sheep were drowned. After the storm, it was said that so much salt lay on the fields of the Isle of Wight that it looked like snow, while sheep shunned the grass of the South Downs until they were desperate. Once they had eaten it, they were so thirsty that they 'drank like fishes'.[29]

In 1703 about half a million people lived in London – one-tenth of the whole population of England. Here, too, churches were damaged, with lead rolled up and carried 'incredible' distances. The Inns of Court and Gresham's College also suffered. Everywhere windows were broken by flying debris. Indeed, according to the *Post Man* newspaper, there was 'hardly any house but has had a share in the calamity'. Londoners faced a cruel dilemma: should they stay inside a house that might collapse or go out and risk being hit by whizzing missiles? A 90-year-old woman from Jewin Street, near the Barbican, who did go out was killed by a brickbat, while a Mr Simpson in

Threadneedle Street resisted pleas to leave his home and was buried under rubble when it collapsed. It is said that as the storm reached an alarming pitch, a carpenter's wife in White Cross Street begged her husband to read a passage from the Bible. He refused and was promptly killed by a falling chimney, while 'miraculously' she was left unhurt. At St James's Palace, Queen Anne is reported to have watched in alarm as trees were torn up in St James's Park, and when chimney stacks toppled and part of the roof fell in, she had to be taken to a cellar for her own safety.[30]

The author Daniel Defoe was narrowly missed by a house that collapsed into the street as he fought his way through the howling winds. If he had been killed, the tempest would have lost its most important chronicler. Defoe himself saw tiles being blown 40 metres, then embedded 20 cm into the ground. After the event, he advertised in the *London Gazette* for witnesses to send in their accounts and used the material to compile a classic account, *The Storm*, published the following year. His overall verdict was: 'No pen could describe it, nor tongue express it, nor thought conceive it unless by one in the extremity of it.'[31]

All over the country, trees were felled: up to 500 at Whitelackington Park in Somerset, more than 4,000 in the New Forest and 3,000 oaks in just one part of the Forest of Dean, along with 'fruit and orchard trees sans number' in the words of the diarist John Evelyn, who himself lost 2,000 trees at his house at Wotton, near Dorking. Philosophically he commented: 'I thank God for what are yet left standing.'[32] With the true journalist's spirit of inquiry, Defoe set out to count how many had been blown down in Kent, but gave up when he got to 17,000. He did, though, calculate that 25 parks had lost more than 1,000 each, and the number lost across the country almost certainly ran into millions.[33]

The death toll on land was probably about 125, but the Great Storm claimed most of its victims on the water. It raged 'furiously' on the Thames, 'breaking a great number of lighters and boats to pieces', so the next morning, the river 'was seen to be all strewn with their wrecks.' Defoe recorded: 'the force of the wind had . . . laid them so upon one another as it were in heaps.' It had been of such ferocity that 'no anchors or land-fast, no cables or moorings would hold them.' He reckoned he saw about 700 wrecked, and hardly any that had escaped without damage. More than 20 people were drowned in London, including two watermen at Blackfriars, and five people on a boat that capsized near Fulham.[34]

But the Great Storm was even more lethal on the waters around our coasts. The days leading up to it had already brought very turbulent weather so that harbours were crammed with ships taking shelter, while others lay at anchor outside ports such as Bristol, Plymouth, Falmouth, Milford Haven and Yarmouth. More than 100 were anchored in the Downs, the area between the east coast of Kent and the Goodwin Sands, where they joined many naval vessels taking a break from the War of the Spanish Succession. Some observers said they had never before seen such a concentration of shipping around our shores. On land, wind speeds had been reaching about 120 km/h (75 mph), but at sea, there may have been gusts of 225 km/h (140 mph). Three ketches went down off Brighton with the loss of all aboard, except for one man who clung to a floating mast for three days. Three merchantmen were sunk off Plymouth; at Milford Haven perhaps 30 ships were lost. At Grimsby almost every vessel in the harbour was blown out to sea, and twenty never returned. The *John and Mary* was sent scurrying 240 km (150 miles) from Great Yarmouth to Scarborough, with the master observing: 'Such a tempest as this, there never was in the world.'[35] Another merchant ship was blown all the way to Norway.

The Navy suffered a terrible toll. The *Newcastle* was driven aground near Portsmouth, with the loss of 197 men. In Yarmouth Roads the *Reserve* was lost with all 190 hands. It was almost impossible for commanders to give orders. According to Defoe, 'words were no sooner uttered than they were carried away by the wind . . . and when they opened their mouths, their breath was almost taken away.'[36] The biggest ship lost in the storm was the 90-gun *Vanguard*, which went down in the Medway, but fortunately there was no one on board. In the Downs the

Prince George had been riding out the storm when her commander saw the *Restoration* bearing down on them. The ships got entangled in spite of the efforts of the crews. For half an hour they rode together until the *Restoration* broke free, only to break up with the loss of all 386 men on board. By dawn a dozen ships had been driven ashore on the Goodwins and other sandbanks. A few hours later, they had all been dashed to pieces and most of their crews drowned – a loss of up to 3,000 lives. The only survivor from the 272 aboard the *Mary* was a seaman named Thomas Atkins, who was washed off her as she broke up and thrown onto the quarter deck of the *Stirling Castle*. When she ran aground, he was flung overboard into a boat which had broken adrift. He reached shore nearly unconscious and suffering from exposure, but alive, along with just 70 others from the crew of 349. The day before the storm struck, it was said the fleet of ships in the Downs had made it look like 'a goodly forest'. Afterwards, it was 'reduced to a desert'. The Royal Navy lost fourteen ships, and overall the death toll at sea was put at 8,000. No wonder the *Observator* newspaper declared: 'never was such a storm of wind, such a hurricane and tempest known in the memory of man.'[37]

5 Literature

'A desert place. Thunder and lightning. Enter three witches.' The opening of Macbeth. Then the famous words: 'When shall we three meet again/ In thunder, lightning, or in rain?' Indeed, throughout the play, thunder is the weird sisters' harbinger. But they also have the power to conjure up a storm. One, angry with a sailor's wife who refuses to give her some chestnuts, gets her two sisters to each provide a wind so the mariner's ship will be 'tempest-tost' (I, iii). But the awesome power of storms is not only associated with the witches in the play. A tempest blows up on the night Macbeth murders the Scottish king, Duncan, with one character talking of 'lamentings heard i' the air; strange screams of death'. Chimney stacks are blown down, and the storm is so fierce that the speaker's 'remembrance cannot parallel a fellow to it' (II, iii).

This notion of a storm being an echo or a warning of a momentous event also provides dramatic scenes in Shakespeare's *Julius Caesar*. On the night before the Ides of March, Casca wanders the streets of Rome with his sword drawn, in a thunderstorm the like of which he has never seen:

> I have seen tempests, when the scolding winds
> Have rived the knotty oaks, and I have seen
> The ambitious ocean swell and rage and foam,
> To be exalted with the threatening clouds:
> But never till to-night, never till now,
> Did I go through a tempest dropping fire. (I, iii)

Casca concludes there must be 'civil strife in heaven' or that humans have been behaving so badly the gods are punishing them. The Roman sees other wonders too, including a lion who 'gazed upon me and went surly by', and concludes that these phenomena are 'portentous' things. He meets Cassius, who, far from being alarmed, declares it 'a very pleasing night to honest men'. When Casca asks him who ever knew the heavens so full of menace, he replies: 'Those that have known the earth so full of faults.' Cassius is angry about the way Julius Caesar is being allowed to lord it over them, and Casca says he has heard rumours that the following day Caesar will be crowned king. At this point Cassius reveals he has persuaded some of the 'noblest-minded Romans' to join him in 'an enterprise of honourable-dangerous consequence' and that the elements are in tune with 'the work we have in hand/ Most bloody, fiery, and most terrible.'

Caesar too has been kept awake by the storm, observing: 'Nor heaven nor earth hath been at peace tonight.' His wife Calpurnia tries to persuade him not to leave the house, because although 'When beggars die, there are no comets seen;/ The heavens themselves blaze forth the death of princes' (II, ii). Caesar scorns

Macbeth, Banquo and the three witches, watercolour from 1876.

King Lear (played by the celebrated David Garrick) rages in the storm. An engraving from 1761.

her fears, and Cassius, Casca and the other conspirators stab him to death at the entrance of the Capitol.

In another of Shakespeare's great tragedies, *King Lear*, the storm mirrors not just the turbulence of the times, but the inner turmoil of the main character. The old king has divided his realm between two of his daughters, Goneril and Regan, each of whom are to take turns playing host to him, along with his 100 followers. But once he has given up his kingdom, his daughters begin to treat Lear with a marked lack of respect, putting his messenger in the stocks and demanding that he get rid of his retinue. In his fury, he flounces out into the storm, urging the elements to mimic his rage at life's injustice:

> Blow, winds, and crack your cheeks! rage! blow!
> You cataracts and hurricanoes, spout
> Till you have drench'd our steeples, drown'd the cocks!
> . . . And thou, all-shaking thunder,
> Smite flat the thick rotundity o' the world! (III, ii)

Two of the select few who have remained faithful to Lear, Kent and the Fool, try to persuade him to take refuge in a hovel, but the king refuses, saying the storm stops him thinking of something worse: 'This tempest will not give me leave to ponder/ On things would hurt me more' (III, iv). He does, however, send his Fool inside, musing that in the past he has taken 'too little care' for those who were less fortunate than him – 'naked wretches' with nowhere to shelter.

The great playwright also uses storms to produce the same kind of sudden changes in fortune that they can cause in real life. In *Pericles* tempests at sea change the direction of the story not once, but twice. It is in Shakespeare's last play, though, that a storm enjoys its most central role. This work is, after all, called *The Tempest*, and is thought to have been inspired by an account the writer read of the wreck of the *Sea Venture* in 1609 on what were then believed to be the enchanted islands of Bermuda. The play opens on the deck of a ship caught in a storm, with those on board in fear for their lives. Then we learn the tempest has been conjured up by a magician, Prospero, who lives on a remote island with his daughter, Miranda. She is distraught at the danger to the mariners and begs her father to calm the weather as the vessel runs aground. Prospero then reveals their back story. He had been the Duke of Milan, but his devotion to learning opened the way for his treacherous brother, Antonio, to depose him with the help of the king of Naples. With Miranda, then aged three, he was put to sea in a leaky old boat, but thanks to 'Providence divine' they found their way to an island inhabited only by the spirit Ariel and the savage Caliban. Then Prospero reveals he raised the storm because he had discovered that his enemies, the usurper Antonio with his accomplice, the king of Naples, were sailing by the island. Now they are marooned and in his power. All manner of strange things happen to the stranded characters. Ariel puts them to sleep or leads them around the island, invisibly playing music. When a banquet is laid out for them, he makes it mysteriously vanish before they can tuck in, then summons up spirits to perform a masque for them. Eventually Prospero forgives Antonio, while the king's son falls in love with Miranda. At the end of the play, the magician discards his magic staff and book and asks the audience to set him free from his exile.

This function of the storm – transporting characters into a strange new world, behaving like a wormhole in space, or the rabbit hole in *Alice's Adventures in Wonderland* – occurs frequently in literature. One of the first novels to use it as a device was *Robinson Crusoe* by Daniel Defoe, whom we last met as the chronicler of England's Great Storm. The book, published in 1719, tells how the young Crusoe defies the wishes of his comfortably off parents and goes to sea. His ship has scarcely set sail before he is caught in a storm. The young man believes he is going to die, and swears to God that he will return to his family if only his life is spared. But once the squall has passed, so does the mood. Crusoe begins to delight in the beauty of the sea, while his shipmates tease him for being so afraid and thinking it such a 'terrible' tempest. Says one: 'do you call that a storm? why, it was nothing at all.'[1] A few days later, though, along comes a storm considered worthy of the name by everyone on board. It sinks the ship off Norfolk, leaving the crew to escape in a small boat. Nothing daunted, Crusoe goes back to sea. This time he is captured by pirates and made a slave. After a couple of years he escapes, and is fortunate enough to be picked up by a Portuguese captain who takes him to Brazil, where the hero becomes a successful plantation owner. In spite of his own unhappy experience of losing his freedom, a few years later Crusoe sets off to West Africa intending to bring back a cargo of slaves. Instead he encounters the storm that will transport him to a very strange place indeed.

At first, his ship hugs the coast 'in very good weather, only excessively hot', but once they begin to cross the ocean, they encounter a hurricane which blows

> in such a terrible manner, that for twelve days together we could do nothing but drive, and, scudding away before it, let it carry us whither fate and the fury of the winds directed; and, during these twelve days, I need not say that I expected every day to be swallowed up.

Having been driven far off course, and sustained damage to their ship, they decide to head for Barbados to get help, but soon they are hit by another storm, which drives the ship aground on a sandbank where 'the sea broke over her in such a manner that we expected we should all have perished immediately.' Fearing the vessel will break up at any moment, they climb into a boat and commend themselves 'to God's mercy and the wild sea'. But 'the sea went so high' that their plight seemed little better than it had been aboard the ship, and after they had covered about five miles, 'a raging wave, mountain-like, came rolling astern of us . . . It took us with such a fury, that it overset the boat at once; and separating us as well from the boat as from one another' so they were 'all swallowed up in a moment'. Crusoe is a strong swimmer, but he relates: 'I could not deliver myself from the waves so as to draw breath.' A particularly powerful wave carries him ashore, 'half dead with the water' he has taken in. He tries to run up the beach to avoid being carried out again by the next wave, but it comes for him 'as high as a great hill, and as furious as an enemy', dragging him perhaps 30 feet down. Desperately

John William Waterhouse, *Miranda – The Tempest*, 1916, oil on canvas. Miranda surveys the ship wrecked by the storm her father conjured up.

Robinson Crusoe and his Man Friday, *c.* 1874, lithograph.

he holds his breath, but just when he is 'ready to burst', the waters carry him ashore once again.[2] This happens several times more, with the waves once dashing him against a rock and knocking him senseless, but eventually he is able to scramble up the beach out of their reach. Soon he realizes that all his comrades have drowned. Crusoe has been shipwrecked on an uninhabited island off the mouth of the Orinoco, where he will be marooned for 28 years, his adventures making up one of the most famous stories in English literature as he builds a shelter, finds food, keeps a calendar, dodges cannibals, rescues his faithful servant Friday, and so on.

A hurricane on land brings a Crusoe-like transformation to the lives of a group of children in Richard Hughes's novel of 1929, *A High Wind in Jamaica*. The Bas-Thorntons are growing up on the island at some unspecified time after the end of slavery. Like many of the grand houses that flourished before emancipation, theirs has now fallen into acute disrepair. The stone ground floor has been given over to goats, while the wooden upper storey occupied by the family leaks like a sieve. But to the children, Jamaica seems 'a kind of paradise', with the three girls wearing their hair short, climbing trees and trapping animals just like their brothers.

One Sunday evening, as a thunderstorm starts, an elderly black man called Lame-foot Sam finds one of Mr Thornton's best pocket handkerchiefs lying on the ground. Because it is the Lord's day, he does not like to just waltz off with it, so he covers it with earth, planning to collect it the next day. But then the temptation becomes too great, and, Sunday or not, he takes the handkerchief. The storm grows stronger as Hughes tell us:

> in the Tropics a thunderstorm is not a remote affair up in the sky, as it is in England, but is all around you: lightning plays ducks and drakes across the water, bounds from tree to tree, bounces about the ground, while the thunder seems to proceed from violent explosions in your own very core.

The children go out to meet their father as he arrives home and see lightning flashing about his stirrup-irons. In seconds, they are drenched.

All through their supper, 'the lightning shone almost without flickering'. While they are eating, Lame-foot Sam comes into the house. Fearing that his pilfering has caused the storm, he flings down the handkerchief. But if the thunder and lightning is the Almighty's retribution, He is not appeased, because soon after, Sam's hut goes up in flames. The old black man starts throwing stones at the sky, protesting that he had given the offending object back, when a blinding flash kills him where he stands. As the wind turns into a hurricane, Mrs Thornton begins reciting the Psalms and the poems of Sir Walter Scott. It must have been a gripping performance, because the children listen with rapt attention even when the shutters burst, and rain pours in 'like the sea into a sinking ship'. The wind snatches pictures from the wall and strips the table bare. Outside, the bushes are blown flat, 'laid back on the ground as close as a rabbit lays back his ears', while 'the negro huts were clean gone'. Exasperated that in the middle of a hurricane, the children have ears only for *The Lady of the Lake*, Mr Thornton shouts that in half an hour they will all probably be dead, but it is only when the roof blows off two cantos later that the family agree to move down

A high wind in Jamaica. An aeroplane upended by Hurricane Gilbert at Kingston airport, September 1988.

to the cellar through a hole he has smashed in the floor. It is crowded with black people and goats already sheltering. Mr Thornton hands out Madeira wine to all and sundry, and the children fall asleep.

The rain stops just before dawn, and the damage is revealed. The wooden part of house has virtually disappeared, with the furniture smashed into matchwood, while the country around is 'quite unrecognisable'. All vegetation for miles has been reduced to 'pulp', and the ground ploughed up by 'instantaneous rivers'. Thornton and his wife look around dazed: 'It seemed not credible that all this had been done by a current of air.' They decide the storm was a warning from heaven, and pack their children

off to England. The youngsters are surprised as, apart from their distress at the loss of a favourite cat, they have been rather unimpressed by the hurricane. In fact, the experiences into which it pitches them will be far more bizarre, because during the voyage to England, they are captured by pirates. After that, they have a series of rather dream-like adventures, as the eldest boy falls to his death from a window during a nativity play, while the eldest girl kills a man.[3]

In Carol Birch's novel *Jamrach's Menagerie*, published in 2011, the young hero is plunged into a strange new world by a much rarer literary phenomenon: a waterspout. Jaffy Brown has diced with death as a child by walking up to a tiger and stroking its nose. When he is fifteen, he joins a perilous expedition to Indonesia to capture a dragon. The beast is successfully taken, but once it is on board, some of the crew believe it is bringing bad luck. One night, it escapes from its cage and plunges into the ocean. Soon after, they encounter a strange sight: 'the dark

A waterspout climbing 'through the sky like the beanstalk in the old story'.

cloud ceiling had a bloated, boiling look, but was luridly bright in one place. From here, a long, white serpent, swaying gracefully, reached down to the surface of the ocean.' As the captain frantically sets the sailors to work in an effort to avoid the waterspout, Jaffy just wants to watch this 'lovely whirling, dreamy thing dancing across the water with furious speed'. He had seen many wonderful sights since he left home, 'but nothing to match this . . . It looked as if the cloud was sucking up the sea through a spinning column of luminous mist.' He notes that the object seems to halt about a mile away as if to watch them. The waterspout 'climbed through the sky like the beanstalk in the old story', while at its foot there was 'a massive commotion'.

Then another waterspout appeared: 'a beautiful oyster-pearl column with what seemed like a pale cloud rising within, and then a third, wider at the top like the funnel of a trumpet and tapering down' to the water. The trio 'danced a stately court dance, three willowy girls weaving in and out of one another, advancing, retreating, bowing and bending, coming together to part and circle, deft and elegant in every move as nymphs and fairies but stronger than Hercules'. And moments later, they demonstrated that power, as the 'most slenderly girlish' ran at the ship 'with the roaring of the deluge and unimaginable speed', raising a wave that rocked them violently. Next came one of her 'sisters' and, 'with the sound of a mountain falling', set the vessel spinning round in circles. But it was the third that dispensed the coup de grâce. This one resembled 'a horn of plenty from which the massed clouds of the canopy had burst forth like foam. The stem was a mighty trunk, grey, shot through with quivers of lightning.' Beautiful she might be, but for Jaffy, she had become a 'living monster' pursuing them. The ship 'tacked about like a flying bat, but she mimicked us, playing Simon Says, turning as we turned, changing as we changed. You'd have sworn there was a brain in that thing.' For a mile, they tried to evade her, but they were 'hopelessly outrun' and she hit them: 'The world rolled round and I rolled with it, banged and winded and beaten by the timbers of the deck.' The mainmast broke 'with a great crack, toppling like a tree, lifted and blown away like a twig by the wind.' Jaffy is one of a dozen of the crew who make it into the boats, where they will remain for weeks, starving and dying of thirst until the hero has to kill, and eat, his best friend, emerging as one of only two survivors.[4]

But literary storms can transform characters' lives without transporting them to exotic corners of the world. 'The Drunkard' is a typically sardonic short story from the nineteenth-century French author Guy de Maupassant. It opens with the north wind 'blowing a hurricane', and behaving in a storm's usual mischievous way: 'whistling and moaning, tearing the shingles from the roofs, smashing the shutters, knocking down the chimneys, rushing through the narrow streets in such gusts that one could walk only by holding on to the walls.' In a little Normandy fishing port a man named Mathurin persuades his friend Jeremie to shelter with him in a tavern, and plies him with drink. When they finally leave, Mathurin disappears swiftly into the night, while, thwarted at every turn by the storm and his own inebriation, Jeremie can barely struggle home. As he eventually stumbles through his door, he thinks he sees a figure rush out past him into the darkness. Convinced that his wife has been illicitly entertaining Mathurin, he beats her to death in a drunken rage.[5]

In 'The Snow Storm', a short story from 1831 by the great Russian writer Alexander Pushkin, a storm plays a more complex trick. Two lovers have decided to elope and marry against their parents' wishes at a country church, but a fearful blizzard breaks out so that when the man sets off, he can see nothing. The road and every landmark has disappeared, and his horse keeps stumbling into snowdrifts or hidden holes. Completely lost, he arrives at the church hours late to find it deserted. Earlier a hussar, trying to rejoin his regiment, has become hopelessly lost in the same blizzard. At last he sees a light, and finds himself outside a little wooden church. In its dimly lit interior are a priest and a number of other people who are relieved at his arrival but keep reproaching him for being so late. 'Impelled by an incomprehensible, unpardonable levity',[6] he lets himself be married there to the agitated young lady waiting at the altar. When the priest tells them they are now man and wife, the lady realises the man who has just become her husband is not her lover and falls senseless to the ground, while the hussar takes to his heels, never knowing who his new wife was or even in which village they were married. Four years later, the hussar and the woman meet again, and fall in love.

Not surprisingly, tempests feature prominently in the literature of the Romantic era, often tuning in with dramatic moments in the characters' lives. So in Emily Brontë's *Wuthering Heights* (1847), a storm 'came rattling over the Heights in full fury' as Cathy wept vehemently and bitterly over Heathcliff's disappearance,[7] while in her sister Charlotte's *Jane Eyre*, also published in 1847, it is when Mr Rochester finally declares his love for the heroine that the weather echoes the violence of human passions. A 'wild wind' roars, a tree 'writhed and groaned'. As rain 'rushed' down, 'a livid, vivid spark leapt out of a cloud . . . and there was a crack, a crash, and a close rattling peal.'[8] The next morning, Jane learns that the great horse-chestnut tree at the bottom of the orchard has been sliced in half by lightning.

In 'The Storm', a short story by the Louisiana writer Kate Chopin, published in 1898, wild weather provides not only an echo of lovers' passions, but the opportunity to satisfy them. The narrative begins with Bobinôt and his four-year-old son, Bibi, in a store. They decide to stay for a while because a storm is brewing, and Bobinôt buys his wife, Calixta, a can of the shrimps she loves. She is at home sewing, and has started to close the windows and doors when a man named Alcée rides up. He asks if he can shelter until the storm passes, and she invites him in. Alcée has not seen much of Calixta since her marriage, but he notes that her blue eyes 'still retained their melting quality'. They are in the sitting room, but the bedroom door is open, and 'the room with its white, monumental bed, its closed shutters, looked dim and mysterious.' It is all beginning to sound rather reminiscent of Dido and Aeneas taking refuge from the tempest in their cave. As the rain clatters down 'in sheets' and lightning flashes, Calixta voices her worries about her husband and son. Then a bolt of lightning strikes a tree with a terrifying crash, and she falls into

Alcée's arms. For an instant he holds her close, then she breaks free, fretting that 'the house'll go next!'

The man grabs her by the shoulders and looks into her face, and we learn that there had once been something between them (though readers of an earlier Kate Chopin story, 'At the 'Cadian Ball', would already be in on the secret): 'The contact of her warm, palpitating body when he had unthinkingly drawn her into his arms, had aroused all the old-time infatuation and desire for her flesh.' He reassures her that the house is too low to be struck. The sight of her face, her hair, her lips, her neck, her breasts 'disturbed him powerfully'. While on her side, 'the fear in her liquid blue eyes had given place to a drowsy gleam that unconsciously betrayed a sensuous desire.' Their lips meet, and he reminds her of a time when 'he had kissed her and kissed her and kissed her.' In the bedroom, 'the generous abundance of her passion, without guile or trickery, was like a white flame.' As they lie in each other's arms, the storm is no longer a threat to them: 'They did not heed the crashing torrents, and the roar of the elements made her laugh.' And their encounter is worthy of the tempest: 'when he possessed her, they seemed to swoon together at the very borderland of life's mystery.' With the end

Storm clouds over a Louisiana cypress swamp.

of their love-making, the storm dies down: 'the growl of the thunder was distant and passing away.' As the sun comes out, Alcée rides away 'with a beaming face' while Calixta laughs out loud. Struggling back across the muddy fields with Bibi, Bobinôt is worried that he will be told off for getting his clothes dirty, but Calixta is sweetness itself and declares they will have a feast when he presents the can of shrimps. Alcée, meanwhile, writes to his wife who is having a short holiday to say it is all right if she wishes to stay away a little longer. Devoted, though she is to her husband, she is delighted at the suggestion. 'So', says the dead-pan final sentence, 'the storm passed and every one was happy.'[9]

Rather than echoing his mood, the Romantic poet Samuel Taylor Coleridge calls on a storm to change it. Tormented by an unhappy marriage and his love for another woman, in 'Dejection: An Ode' (1802), he lies in 'a stifled, drowsy, unimpassioned grief', wishing the wind would get up and 'startle this dull pain, and make it move and live!' Before the end of the poem he gets his wish, as the tempest breaks, emitting the sound of 'the rushing of a host in rout,/ With groans, of trampled men, with smarting wounds', then reminding him of

> . . . a little child
> Upon a lonesome wild,
> Not far from home, but she has lost her way:
> And now moans low in bitter grief and fear,
> And now screams loud, and hopes to make her mother hear.

Many English poets have written evocatively about storms – John Clare, Wilfred Owen, Siegfried Sassoon and James Thomson, to mention just a few. But perhaps not surprisingly, in view of the more extreme weather experienced across the Atlantic, it is often Americans who come up with the most vivid lines. In 'A Thunderstorm' Emily Dickinson (1830–1886) describes lightning that 'showed a yellow Beak/ And then a livid Claw.'[10] In 'The Snow-Storm' Ralph Waldo Emerson (1803–1882) told of individuals stranded in the 'tumultuous privacy of storm', while in 'Hurricane' the Jamaican-born poet James Berry

(b. 1924) captured the terror of a hurricane: 'speedy feet, all horns and breath,/ all bangs, howls, rattles', as it tore down roofs, felled the toughest trees and battered the fields. But the poet is also struck by the odd phenomena it generated – sending zinc sheets flying like kites and leaving dead fish lying in the road. While Berry concentrates on the dramatic effect of the hurricane on helpless humans, for the nineteenth-century American poet William Cullen Bryant, the tropical storm is the manifestation of God's power and glory, and in 'The Hurricane' from 1854 he awaits it 'with a thrill in every vein'. He describes how it sails 'Through the boundless arch of heaven . . . / Silent and slow, and terribly strong'. Its 'mighty shadow' is a reminder of 'the dark eternity to come'.

Specific historical storms also feature in literature. The English Jesuit poet Gerard Manley Hopkins wrote 'The Wreck of the Deutschland' in 'happy memory' of five Franciscan nuns who were killed when their ship ran aground off Harwich during a blizzard in 1875,[11] while the Great Storm of 1987 plays a crucial role at the climax of A. S. Byatt's *Possession*, winner of the Booker Prize in 1990. The story begins in a London library, where a rather unsuccessful English literature scholar named Roland Mitchell discovers a letter written by a fictional nineteenth-century poet, Randolph Henry Ash, which leads him to believe that Ash may have had a hitherto unknown love affair with another fictional poet, Christabel LaMotte. The great expert on LaMotte is a distant relative of hers, Dr Maud Bailey. Mitchell and Bailey become obsessed with discovering the truth, and a close and sexually charged though unconsummated relationship develops between them. But they face fierce academic competition from an unscrupulous American, Mortimer Cropper, who wants to corner the market in all things Ash.

Cropper decides that proof of Ash's affair with Christabel may lie in a box of documents buried with the poet, and manages to persuade one of his descendants, Hildebrand Ash, to dig it up. At one o'clock on the morning of 16 October 1987, the pair head for the Sussex country churchyard that is the poet's last resting place. It is raining a little, while the air is 'heavily still'. As they start work, the wind begins to get up. Trees creak and moan. Bewildered, the American stops digging, and 'in that moment, the great storm hit Sussex.' Hildebrand is winded by 'a wall of air' while Cropper soon gets shovelling again. Now there is a 'chorus of groans' as the trees begin gesticulating desperately. A tile flies off the church roof, while the wind becomes 'like a creature from another dimension, trapped and screaming'.

Hildebrand is scared, and says they must take shelter, but, with the trees swinging about ever more ominously, Cropper finds the box. By now the tiles are a threat to life and limb, zipping through the air and crashing into gravestones. As the American tries to put his treasure in the boot of his Mercedes, 'a great mass of grey descended before his eyes like a tumbling hill.' His car is crushed by a fallen tree. Then he hears voices, and a shout: 'You're surrounded!' Mitchell, Bailey and their friends take possession of the box, which they claim rightfully belongs to Bailey, who, it transpires, far from being a distant relative of LaMotte, is actually directly descended from an unknown illegitimate daughter she had by Randolph Henry Ash. The group manage to cut their way through fallen trees to a local inn, crowded with other refugees from the storm. There they open the box and find the love letters exchanged by the two poets, and that night Mitchell and Bailey finally become lovers. The morning after the storm, writes Byatt, 'the whole world had a strange new smell. It was the smell of the aftermath . . . It was the smell of death and destruction and it smelled fresh and lively and hopeful.'[12]

Mighty though literary storms can be, sometimes a character appears who is foolish enough to believe he is even more powerful, like the captain in Henry Wadsworth Longfellow's famous poem 'The Wreck of the *Hesperus*' (1841). One of the ship's crew warns the skipper that a hurricane is on the way, and begs him to put into port, but his commander lets out a 'scornful laugh'. Then the wind hits the vessel, which 'shuddered and paused, like a frighted steed.' Unfortunately, the captain has chosen to take his daughter on this voyage, but he tells her not to be afraid because he can weather 'the roughest gale/ That ever wind did blow.' Then he wraps her in a warm coat and ties her to the mast. As the storm rages, she cries out three times to her father. Twice he replies, but the third time there is silence. The sleet and snow has reduced him to a 'frozen corpse'. The girl thinks of how Christ stilled the waves on the Sea of Galilee, and prays she may be saved as the *Hesperus* charges on, but in the end, 'cruel rocks' gore the ship 'like the horns of an angry bull', and, covered in ice 'like a vessel of glass', she sinks. The next day, a fisherman finds the poor girl dead, still lashed to a floating mast. Longfellow's poem is thought to have been inspired by accounts of a great blizzard off the east coast of America in 1839.

In Joseph Conrad's 1903 classic *Typhoon*, another captain, Tom MacWhirr, refuses to be diverted by a storm. He can see from the barometer that there must be 'some uncommonly dirty weather knocking about', but when his first mate, Jukes, urges him to change course to avoid it, he refuses, saying there was never any shortage of bad weather, 'and the proper thing is to go through it'.[13] Conrad, who had been a seaman himself, and had experienced being shipwrecked, wrote that in all his years at sea, MacWhirr 'had never been given a glimpse of immeasurable strength and of immoderate wrath . . . the wrath and fury of the passionate sea.' He knew it existed, but only in a theoretical way: 'he had heard of it as a peaceable citizen in a town hears of battles, famines and floods.' The captain 'had sailed over the surface of the oceans as some men go skimming over the years of existence to sink gently into a placid grave, ignorant of life to the last . . . There are on sea and land such men thus fortunate – or thus disdained by destiny or by the sea.'[14]

MacWhirr seems a pretty dull old stick. His son and daughter scarcely know him because of his long absences, while his wife lives in 'abject terror' of her husband giving up seafaring and coming to live with them.[15] But the sea is to disdain the captain no longer once he takes command of a steamer, the *Nan-Shan*, carrying 200 'coolies' back to their homes in China. From what he has learned of MacWhirr so far, the reader is surely entitled to conclude that his encounter with an approaching typhoon is likely to end badly – a suspicion that is reinforced

Thomas Ender, *The Steamship 'Marianne' in a Storm on the Black Sea*, *c.* 1837, oil on canvas.

when the storm strikes while the skipper is asleep. He wakes to see his shoes 'scurrying from end to end of the cabin, gambolling playfully over each other like puppies'. When he goes on deck, he at once finds himself 'engaged with the wind in a sort of personal scuffle'. Looking around, all he can see is 'a great darkness lying upon a multitude of white flashes', while the wind has taken 'upon itself the accumulated impetus of an avalanche'. The ship begins 'to jerk and plunge as though she had gone mad with fright'.

As MacWhirr stares 'into the wind's eye as if into the eye of an adversary, to penetrate the hidden intention and guess the aim and force of the thrust', the storm gets even worse, like 'the sudden smashing of a vial of wrath. It seemed to explode all round the ship with an overpowering concussion and a sudden rush of great waters, as if an immense dam had been blown up.' Unlike other natural disasters – earthquakes, landslides, avalanches – writes Conrad, a storm is 'like a personal enemy'. Now

Sailors high in the rigging as the *Garthsnaid* battles the sea, *c.* 1920.

it seemed 'the whole China Sea had climbed on the bridge', and the mate Jukes began to despair. The typhoon had become so violent, it 'appeared incompatible with the existence of any ship whatsoever'. The *Nan-Shan*'s 'lurches had an appalling helplessness: she pitched as if taking a header into a void, and seemed to find a wall to hit every time', while the wind's gusts were so fierce they appeared to raise the vessel right out of the water. Jukes and the captain clung to each other as the sea lifted them up and then flung them down 'with a brutal bang'. Equipment was torn from the decks as if the ship was being 'looted by the storm with a senseless, destructive fury', leaving the *Nan-Shan* 'like a living creature thrown to the rage of a mob'. Jukes keeps having to inform the captain of some fresh misfortune, such as the boats being lost, and MacWhirr always shouts back in the same laconic, matter-of-fact tone: 'All right' or 'Can't be helped.' The voice, writes Conrad, has 'a penetrating effect of quietness in the enormous discord of noises', like one 'that shall be pronouncing confident words on the last day, when the heavens fall'.

And speaking of heavens falling, approaching them now is 'a white line of foam coming on at such a height' that there can be no doubt about 'the awful depth of the hollow the hurricane had scooped out behind'. The ship 'lifted her bows and leaped'. Then, 'with a tearing crash and a swirling, raving tumult, tons of water fell on the deck'. The *Nan-Shan* 'dipped into the hollow straight down, as if going over the edge of the world'. She managed to rise again slowly, 'staggering, as if she had to lift a mountain with her bows'. The steering gear had been torn away, and the chief engineer observed: 'Another one like this, and that's the last of her.' Then mercifully they find themselves in the eye of the typhoon, that brief respite before the storm will come again. MacWhirr gathers his men and tells them: 'Don't you be put out by anything. Keep her facing it.' The last words he utters in the story are another heroic understatement: he 'wouldn't like to lose' the ship. And MacWhirr's strategy seems to work, because the next we know, the *Nan-Shan* is limping into port, looking as though she has been used for target practice, but with no injuries worse than a few broken bones among crew and passengers.

The captain writes an account of his battle with the typhoon for his wife, but she finds it so boring that she does not even notice that for two hours, her husband believed his ship was about to be sunk. The chief engineer writes to his wife that although MacWhirr was 'a rather simple man', he had just 'done something rather clever', though he does not tell her what. And the last word in the story is left to Jukes, whose verdict on the captain is: 'he got out of it very well for such a stupid man.'[16]

6 Spectacle

One of the earliest major paintings to feature a storm is also one of the most enigmatic. Indeed, *The Tempest*, painted by the Italian Renaissance master Giorgione in about 1506, goes by other names too, such as *The Soldier and the Gipsy*. On the right of the canvas sits a woman feeding her baby, naked apart from a shawl across her shoulders. On the left stands a man holding a staff. There is no interaction between the pair, and, it is not even clear whether either is aware of the other's presence. Close to the man are a couple of broken columns, and in the distance behind the figures is a small town over which dark storm clouds have gathered, while a fork of lighting slices the air. Byron thought it the most delightful picture he had ever seen, celebrating the woman as

> Love in full life and length, not love ideal,
> No, nor ideal beauty, that fine name,
> But something better still, so very real.[1]

But what does the picture mean? Some have sought explanations in the Bible, some in mythology, some in allegory, while others, pointing to its dreamlike quality, have asked whether it means anything at all. One thing is clear, though: this is a pretty benign tempest. There is no sign of rain or wind, and neither figure seems to feel any need to take shelter. The juxtaposition of the clothed man and the nude woman was very unusual, and some scholars have seen it as an inspiration for Manet's *Le*

Giorgione, *The Tempest*, *c.* 1506, oil on canvas.

Déjeuner sur l'herbe (1863), which caused a scandal more than three centuries later because of its depiction of a naked woman sitting with two fully dressed men.

The origins of the most famous storm picture by the great Rembrandt are much clearer. His *Storm on the Sea of Galilee* from 1633 is inspired by the well-known Bible story. It shows one of the disciples apparently being sick over the side of the small sailing ship; a couple more seem to be haranguing Jesus and another appears transfixed by fear, while others struggle with the sails. Only Christ Himself looks calm. We know that He is going to still the storm, but things certainly seem dangerous enough at the moment the artist depicts, with the vessel forced up to a crazy angle by foaming, mountainous waves. Or perhaps that should be 'depicted', because the painting was stolen from a museum in Boston in 1990 and has not been seen since.

Long before fine art discovered storms, they had been portrayed in more utilitarian works known as ex-voto, or votive, pictures. These were designed to procure, or give thanks for, a divine helping hand. The tradition went back a long way. A story from ancient Greece relates that one day at a sanctuary to the gods on the island of Samothrace, a man tried to persuade his sceptical friend that the deities did care what happened to humans, and to prove his point gestured around the sanctuary at the 'many painted votive tablets' thanks to which sailors had 'escaped the violence of storms and arrived safely into port'.[2] By Rembrandt's time, a lot of votive pictures had settled into a formula in which the lower half would show the storm at sea from which the donor had escaped, while the upper half featured the saintly figure whose intervention had saved them. The tradition crossed the Atlantic, so the church of Sainte-Anne de Beaupré in Quebec, for example, housed one referring to a tempest in 1709 that was threatening to wreck the barque *Sainte Anne*. A priest sank to his knees on the deck and earnestly beseeched the Lord to save them. Everyone escaped, and to give thanks, the captain and crew had a canvas painted depicting the ship, the stormy sea and the praying priest. Other early storm paintings from North America had an equally utilitarian though different

Rembrandt, *The Storm on the Sea of Galilee*, 1633, oil on canvas.

A 17th-century votive painting from a church at Ortisei in Italy.

'The schooner *Baltick* in distress in 6 fathoms of Water at Nantucket Sholes with everything wash'd of the Decks & Two men Drounded. The 19th of Dec.' Artist unknown, mid-18th century; watercolour on paper.

purpose, being produced by the owners or insurers of vessels to show that everything possible had been done to prevent a shipwreck, and that the disaster was an act of God and not the result of negligence on the part of the crew. One such was an undated watercolour showing the schooner *Baltic* 'on Nantucket Shoals with everything washed off the deck and two men drowned'.[3]

While Rembrandt was creating *The Storm on the Sea of Galilee*, landscape painting was beginning to emerge for the first time as a respected genre. One of its leading lights was the great French master Nicolas Poussin. Most of his pictures featured historical, mythological or biblical works, and in the earlier ones, such as *Et in Arcadia ego* (1637–8), the landscape usually played a supporting role to the human figures. Later it began to steal the show. In his work from 1651, *Landscape During a Thunderstorm with Pyramus and Thisbe*, the tempest provides a fitting setting for the story of the doomed lovers from ancient Babylon. With their parents forbidding their union, they try to flee the city, only for each to commit suicide after a tragic misunderstanding. The human protagonists are tiny, and dwarfed by trees bending in the fierce wind and a black sky riven by forks of lightning. Poussin wrote that he had tried to imitate

> the effect of a sudden and violent wind, an air filled with darkness and rain, with lightning and thunderbolts . . . Every

Nicolas Poussin, *Landscape During a Thunderstorm with Pyramus and Thisbe*, 1651, oil on canvas.

Ludolf Bakhuizen, *Warships in a Heavy Storm*, *c.* 1695, oil on canvas.

> figure seen there acts in accordance with the weather: some flee through the dust, following the wind that carries them; others, on the contrary, go against the wind, and walk with difficulty, putting their hands in front of their eyes . . . Amid this disorder, the dust rises in large vortexes.[4]

When the painter embarked on a series of paintings of the four seasons, for *Winter* (1664) he chose to depict Noah's flood in what may have been his final work. As what are perhaps the last humans left outside Noah's charmed circle struggle to clamber into boats, again the sky is covered by angry clouds and cleft by lightning.

During the seventeenth century in the Netherlands, land- and seascape painters such as Simon de Vlieger and Ludolf Bakhuizen began to depict naturalistic storm scenes, as did other artists like the Italian Francesco Guardi, better known for his serene views of Venice. Paintings from this era often included an area of blue sky – a hint of kinder weather to come and a reminder that storms are part of the natural order.

But by the end of the eighteenth century, a darker mood began to emerge, with painters such as the Frenchman Claude-Joseph Vernet using very stark contrasts of light and dark, against which tiny human figures struck anguished poses. And for the Romantic painters, of course, storms were a godsend. After all, one of the forerunners of Romanticism was a German literary movement known as *Sturm und Drang* – storm and stress – and soon, all over Europe, art was turning away from discipline, reason and convention and towards emotion and passion. Romantic works, rather than stressing the transient nature of a storm, would highlight the impotence of man before the awesome power of nature. Perhaps the greatest of all Romantic painters, J.M.W. Turner, though heavily influenced by the Dutch tradition, would put human figures caught in a desperate struggle right at the centre of his earlier storm pictures, such as *The Wreck of a Transport Ship* from 1810. A seasoned admiral was awestruck by the work, exclaiming: 'No ship or boat could live in such a sea.'[5] Turner became highly successful, but when other

Claude-Joseph Vernet, *Shipwreck in a Storm*, 1770, oil on canvas.

marine painters, such as Augustus Callcott and George Philip Reinagle, emerged to challenge his pre-eminence, the master, never shy of competition, took his art in a new direction.

In 1812 his *Snow Storm: Hannibal Crossing the Alps* adopted an innovative composition based on a dramatic vortex of wind and snow beneath a huge black cloud to capture the ordeal of Hannibal and his troops battling the terrain, the weather and the local tribes. Human figures are to be found only at the foot of the picture, and are completely dwarfed by dreadful natural forces. In his final years Turner's storm pictures became more and more abstract. In *Rain, Steam and Speed – The Great Western Railway*, first exhibited in 1844, we glimpse the locomotive's form and the carriages behind, hints of two bridges and even a little boat on the river, but most of the canvas is occupied by the steam and the swirling rain. In *Snowstorm – Steam-boat off a Harbour's Mouth* the boat is, if anything, even less distinct, and the spiralling vortex of filthy weather conjured up by nature is demonstrating just who is boss to one of the most modern ships

J.M.W. Turner, *The Wreck of a Transport Ship*, *c.* 1810, oil on canvas.

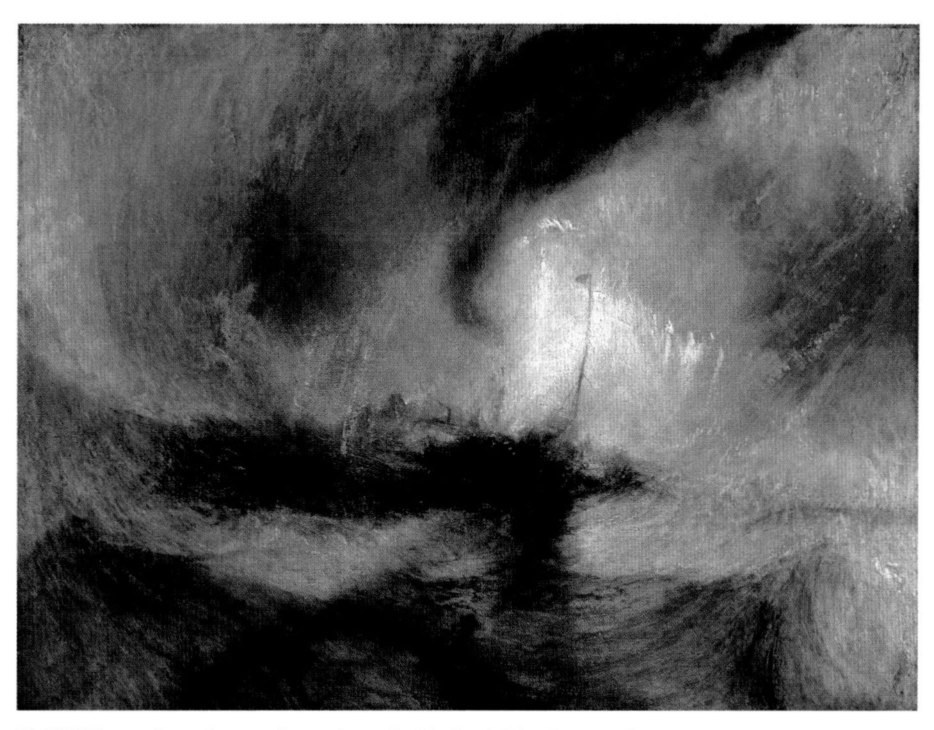

J.M.W. Turner, *Snow Storm – Steam-boat off a Harbour's Mouth*, 1842, oil on canvas.

John Constable, *Stormy Sea, Brighton, 20 July 1828*, 1828, oil on canvas.

on the seas. It is said that Turner once had himself lashed to the mast of a ship so he could observe a tempest at sea, and a similar story is told of Vernet. Turner's great British contemporary John Constable also created storm pictures in which the approach veers towards abstraction, though they are less stylized than Turner's paintings. In works such as *Stormy Sea, Brighton* (1828), conveyed in free, uninhibited brushstrokes, the human element is omitted, and the sea almost merges with the sky.

Many American painters were influenced by Turner and Constable, and they also had the more extreme storms of the New World to inspire them. Jasper Cropsey, who had seen paintings by the English masters on visits across the Atlantic, declared that storms were capable of 'awakening the deepest emotions of gloom, dread, and fear; or sending thrilling sensations of joy and gladness through our being'.[6] Thomas Cole, a founder of the Hudson River School, met Constable in London. When he painted a tornado in 1835 he adopted a straightforward, naturalistic approach, but later he became more interested in storms as a metaphor. In the final picture of his series, *The Course of Empire* (1836), a Turner-like vortex of clouds gathers over a great city as it is destroyed. Cole noted: 'A savage enemy has entered the city. A fierce tempest is raging.'[7] Similarly, in his *Voyage of Life* series (1842), *Childhood* and *Youth* have calm skies, while *Manhood* has storms. The glowering skies in Emanuel Leutze's famous heroic painting from 1851, *Washington Crossing the Delaware*, are seen by some as symbolizing the darkest days of the American revolutionaries, from which they are now emerging as the general stoutly thrusts out his chest while his craft negotiates the icy waters. Similarly, Martin Johnson Heade's powerful series of storm pictures, which appeared at the end of the decade, are often interpreted as a sign of the growing tensions that would soon erupt into the American Civil War.

Jasper Cropsey, *Cove: A Storm Scene in the Catskill Mountains*, 1851, oil on canvas.

Thomas Cole, *The Voyage of Life: Manhood*, 1842, oil on canvas.

The Impressionists too were great admirers of Turner. In October 1886 Monet travelled to Belle Île off the coast of Brittany. The weather was pretty bleak, with a tempest raging for three days, and the painter remarked: 'it is extraordinary to see the sea; what a spectacle! It is so wild you cannot help but wonder whether it can ever regain its calm.'[8] Monet said it was a 'joy' to watch, and pronounced himself distraught when the storm subsided. The 'devilish' sea, though, was 'terrible to render'. Being an Impressionist, of course, he was out there in the worst of the weather, and bemused locals watched him at work: 'booted, covered with woollens, enveloped by an "oilskin" with a hood. The squalls sometimes rip the palette and brushes from his hands. No matter, the painter holds on and goes off to work as to a battle.' He was on a mission to capture the 'fleeting splendours of the storm'. And from his travails, physical and artistic, emerged a masterpiece, *Storm Off the Coast of Belle-Île*.[9]

Those painters who followed the Impressionists often interpreted storms via the various avant-garde schools that went coursing through art as the twentieth century dawned. In 1893 that great portrayer of modern angst, the Norwegian Expressionist Edvard Munch, painted *The Storm* at the coastal village of Åsgårdstrand. From the rocks in the foreground, we can deduce we are by the sea, but we do not see the water. In the background

Emanuel Leutze, *Washington Crossing the Delaware*, 1851, oil on canvas.

Martin Johnson Heade, *Approaching Thunder Storm*, 1859, oil on canvas.

Monet, *Storm Off the Coast of Belle-Île*, 1886, oil on canvas.

welcoming lights blaze in a house, but in front of it a tree is bent almost double by the wind, which also blows the hair of a group of women who are holding their hands to their faces. Are they fishermen's wives desperately afraid for their husbands? During the following decade the great Austrian painter Gustav Klimt produced a wonderfully ominous deserted landscape entitled *Approaching Thunderstorm* with a pointillist poplar tree standing in harm's way. The tree, at Litzlberg, survived on this occasion, but would be felled by lightning in 1928. Around the same time, the Fauve Raoul Dufy painted a typically joyous work, *Storm at Sainte Adresse*, with matchstick figures battling the wind and ships and yachts leaning at crazy angles against the water, all captured before a curving horizon in exuberant colours. The great master of the naive Henri Rousseau produced *Surprised!*, a marvellously idiosyncratic portrayal of a tiger in a jungle waiting to pounce on an unseen prey as the rain buckets down and lightning carves open the sky.

In nineteenth-century America Thomas Cole might have painted a tornado in fairly conventional style, but in 1929 the Regionalist John Steuart Curry gave an almost comic-book look to *Tornado over Kansas*. According to Curry's widow, he had never actually seen a tornado, but anyone who had grown up, like him, in the Sunflower State would have heard plenty of accounts from people who had encountered them, and seen many pictures of the devastation they cause. The homestead from which a family is fleeing, carrying a cat, a dog and a baby, is said to be similar to the one in which the artist grew up. With its portrayal of a muscular, square-jawed husband alongside a pale, frightened wife, the picture was derided by modernist critics, but Curry saw himself as part of a movement to found a new American style of painting that would be distinct from the fads sweeping Europe and based on the traditional values of the pioneers and the frontier. Other artists who took advantage of the Americas' extreme

Edvard Munch, *The Storm*, 1893, oil on canvas.

Gustav Klimt, *Approaching Thunderstorm* (*The Large Poplar II*), 1903.

weather included Winslow Homer, painter of many dramatic storms at sea, who produced watercolours of a hurricane and its devastating aftermath when he was in the Bahamas during the last years of the nineteenth century. The U.S. Impressionist Childe Hassam created a number of atmospheric studies of the fierce snowstorms that frequently grip New York City, while his contemporary John Sloan produced a lively portrayal of New Yorkers fleeing a dust storm on Fifth Avenue.

When the movies arrived, storms provided a great opportunity for special effects. One notable early effort was *The Hurricane*,

Winslow Homer, *Hurricane, Bahamas*, 1898, watercolour and graphite on off-white wove paper.

Henri Rousseau,
Surprised!, 1891,
oil on canvas.

Childe Hassam, *Snow Storm, Fifth Avenue*, 1907, oil on canvas.

made by the great American director John Ford in 1937, which won an Oscar for best sound. A fifteen-minute tropical storm sequence forms the climax of the story of how an unfeeling French colonial regime on a South Sea island persecutes a free-spirited, golden-hearted native man. Trees are uprooted, houses demolished and great chunks of debris hurled at the shore, and finally the island's church is destroyed, all accompanied by the demented howling of the wind, a blinding, swirling rain and screams of human terror. For the *New York Times* critic, it was 'a hurricane to blast you from the orchestra pit to the first mezzanine', created by a special effects director who had already cut his teeth on earthquakes and plagues of locusts.[10] During the tempest the martinet of a French governor wants to set off on a schooner to pursue the hero, who has escaped from prison, but a kindly doctor warns him that if he does, he will 'hear God howl and laugh' at him. And there is a feeling that the hurricane has an element of, if not divine, then poetic justice – that the arrogant colonials had it coming. If so, it is an imperfect justice, for the French governor and his wife survive while most of the natives are killed.

Nearly 70 years later, the special effects team for Roland Emmerich's highly successful disaster movie *The Day after Tomorrow* (2004) must have had a very full schedule. As the earth's climate goes crazy, they have to conjure up a snowstorm in Delhi, then hailstones the size of grapefruit in Tokyo. Soon the strongest hurricane ever recorded is raging – though mostly off-screen. But a team of tornadoes do put in an appearance as they devastate Los Angeles, flinging cars through the air, ripping buildings apart, felling people and even erasing the famous Hollywood sign. There are blizzards in Scotland as the temperature plummets to –150°C and a fearful rainstorm in New York City that floods the streets waist-deep in water and then generates a huge wave half a dozen storeys high. Next on the agenda is the mother of all snowstorms. We learn that all these phenomena are the product of the way melting polar ice has diverted ocean currents, and that they are coalescing to produce a huge global storm, which will 'change the face of our planet'. That prediction certainly seems spot on when the whole of the earth becomes invisible from space because of three great hurricanes. Each has an eye holding supercooled air that instantly freezes anything with which it comes into contact.

The great and the good of the United States, until then rather dismissive of climate change, are now beginning to sit up and take notice, just in time for some spectacularly bad news. The megastorm will last for up to

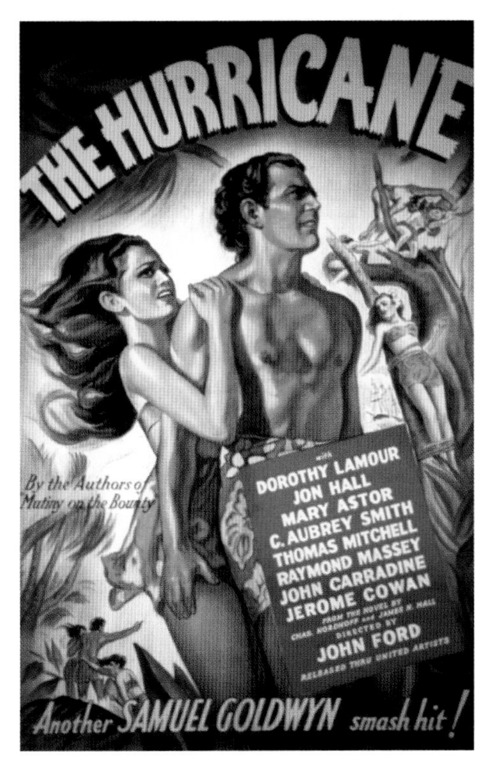

Poster for *The Hurricane*, directed by John Ford (1937).

ten days, covering the whole of the northern hemisphere with ice and snow and sending temperatures plummeting to what they were during the last Ice Age. The people in the southern United States will need to be evacuated. Nothing can be done for those in the north. In a nice twist, the Mexicans close the border as thousands of Americans try to cross illegally, and open up again only when the U.S. president agrees to write off all Latin American debt. Musing on whether mankind will survive, the hero, a climatologist, decides we can if we learn from our mistakes, while the president goes on television to apologize for his past errors in assuming we could just go on consuming the world's resources willy-nilly, and to thank the developing world countries who are now providing homes for American refugees. The film won a BAFTA for its special effects, and was also chosen as the best feature film of the year by America's Environmental Media Association. The critics tended to agree with the *Chicago Sun-Times* that the special effects were 'stupendous', though the same newspaper also lambasted the film for 'cornball plotting',[11] while the *New Yorker* thought it might actually damage the environmental cause: 'the whole project is so dumb, ill-written, and condescending that it may become counterproductive, with viewers fleeing the cinema and vowing never to recycle again.'[12]

The storm as apocalypse, though at a more domestic level, also features in the Coen Brothers' wry black comedy *A Serious Man* (2009), but it makes few demands on the special effects experts. The hero is an American Jewish physics professor, Larry Gopnik,

The Day After Tomorrow (2004): hailstones like grapefruit pulverize Tokyo.

A Serious Man (2009): the tornado approaches.

who finds his life full of troubles. His wife announces she wants a divorce, and makes him move into a motel while she cleans out their bank account. Anonymous letters are being sent to the university denouncing him, his lawyer drops dead when he is on the point of delivering a crucial piece of advice, his brother is facing charges of solicitation and sodomy, and the rabbis to whom Larry turns for advice are unclear, unhelpful or unavailable. Suddenly there is a break in the metaphorical clouds. His boss at the university hints that Larry's job is safe. He is not going to be fired. For some time, the hero has been struggling with a dilemma. Should he accept a large bribe to pass a student he was intending to fail? Larry decides to go for it, and changes the boy's mark. Immediately, the telephone rings. It is his doctor wanting to see him at once about the results of a chest X-ray. At the same moment, Larry's son is being evacuated with fellow pupils from his school to a tornado shelter, but the teacher is having trouble opening the door. As they all stand outside waiting, the wind is really starting to get up, while the ominous dark cloud with a tendril reaching down to the earth gets nearer and nearer. Then the credits roll.

So what does it all mean? That the bad news Larry is about to receive from his doctor will put all his other troubles into perspective? And that the greater disaster of the tornado will make even dire medical tidings seem less important? Or is the whole thing just a shaggy dog story, packed with bizarre, unrelated incidents?

At the start of the 1939 film *The Wizard of Oz*, sixteen-year-old Judy Garland sings wistfully and unforgettably of a faraway land over the rainbow that she once heard of in a lullaby. Judy's character, Dorothy Gale (geddit!?), is upset because her dog has bitten an unpleasant neighbour and now faces being put down. The girl wishes she could go to a place where troubles melt like lemon drops. Minutes later she is being whisked up into the air from her Kansas home by a tornado. The film morphs into colour as Dorothy finds herself in a land inhabited by exotic creatures like the Scarecrow, the Tin Man, the Cowardly Lion and a wicked witch who wants to kill her, and looks uncannily like the neighbour her dog bit. But our heroine becomes more and more desperate to get home, and finally a good witch tells her how. She just has to keep repeating the magic words: 'There's no place like home.' And it works! Dorothy finds herself back on her bed surrounded by her friends and loved ones. She had been knocked out by a windowpane blown in by the storm, and the adventures were all just a dream. The girl vows never to leave Kansas again.

This use of a storm to transport a character into a different world is a stock-in-trade of the movies, just as it has been for the novel. Take *Life of Pi*, adapted by award-winning director Ang Lee in 2012 from Yann Martel's novel. There cannot be many films in which a tempest leaves a boy marooned in a lifeboat in the middle of an ocean with a tiger, and there is surely no other in which the tiger's name is Richard Parker. Pi's parents run a zoo in India, but they run into difficulties with the local council who own the land where it is sited, so his father decides to transport the animals across the Pacific to sell them in Canada, using the proceeds to start a new life. Four days out of Manila, the ship is moving with the 'slow, massive confidence of a continent'. As she passes over the deepest ocean trench on earth, we see her bow breasting huge waves as rain lashes down. Deep below in his family's cabin, Pi hears the distant sounds of the storm, and decides to go up to watch. At first he is exhilarated by the spectacle as he negotiates the crazily slanting decks. Then a great wave surges over the ship. As he tries to get back to the cabin to raise the alarm, he finds himself being washed along a flooded passageway with a zebra. Forced back on deck, he sees some of the zoo's birds and animals flailing around in the tempest. The crew fling him into a lifeboat, and as it is lowered, the zebra jumps in, knocking a sailor into the sea. Huge waves send the boat tumbling over and over, but Pi manages to hang on. In the distance, he sees the ship's lights disappear slowly beneath the ocean's surface, and realizes he is alone, at least as far as human company is concerned. There follows a fight for survival against the elements, but in some ways they are the least of his problems. Pi has to share his boat with the injured zebra, an orang-utan and a hyena, which kills the other two animals, and exhibits little good will towards the hero.

The hyena problem is solved when the tiger Richard Parker suddenly leaps from under the boat's tarpaulin and disposes of it, but Pi then has to find a way of sharing a tiny boat with a very hungry tiger. He survives through a variety of expedients including moving out into a small craft that he improvises from lifebelts and oars. He and the tiger both go catching fish, and Pi even creates a makeshift ladder so the animal can get back into the boat after his expeditions. After months at sea, when they both seem to be on the point of death, they finally reach land, and Pi is deeply hurt to see the tiger just wander off into the jungle without so much as a backward glance at him.

The movie storm can also be a device for keeping a group of characters claustrophobically confined in a pressure-cooker

Life of Pi (2012): owing to a devastating storm, a boy has to share a boat with a tiger.

Prisoners of the hurricane: the tension mounts in *Key Largo* (1948).

environment, as in John Huston's tense thriller from 1948, *Key Largo*. A former army major, Frank McCloud, played by Humphrey Bogart, shows up at a hotel on the Florida Keys run by the father and the widow of one of his soldiers, to tell them about the man's last moments. Already staying is a group of gangsters, led by the villainous Johnny Rocco (Edward G. Robinson), who have dropped in from Cuba with a stash of forged currency they want to sell to another bunch of unsavoury characters. Their plan faces derailment when a hurricane imprisons everyone in the hotel. While Rocco mops his brow and generally gets agitated, his getaway boat disappears, and his buyers are kept away by the storm. Then the hotel owner starts winding him up about a hurricane a couple of miles away in 1935 that swept 800 people out to sea. Getting into his stride, he recounts how winds reached 200 miles an hour and a whole town was destroyed. He then prays aloud that the present storm will kill Rocco. McCloud taunts the gangster too, telling him to threaten the hurricane with his gun: 'If it doesn't stop, shoot it.' While the waves crash and the winds uproot trees, the gangsters murder a policeman, and Rocco is so nasty to his alcoholic girlfriend that she slips a gun to McCloud, which he later uses to kill the villains when they make him take them to Cuba on a boat. As the news that McCloud is safe reaches the widow, the sun comes streaming in through the hotel windows.

Just because a film is packed with special effects does not mean it cannot tell a simple, affecting human tale. Take *The Perfect Storm*, released in 2000. 'Based on a true story', declare the opening titles, though the family of the main character, fishing boat skipper Billy Tyne (played by George Clooney), tried unsuccessfully to sue the film-makers over what they complained was an inaccurate portrayal of him. The film opens with a memorial at City Hall in Gloucester, Massachusetts, which lists the names of those who have died at sea over many decades. Storms are evidently nothing new here. Tyne has just sailed in, and has been getting grief from the owner of his vessel, the *Andrea Gail* (that name again), over his poor run of catches, while his men are disappointed at their pay. To try and change their luck, he persuades his crew to head for the faraway Flemish Cap, and soon the fish are leaping onto their boat. It is a bumper catch, but their ice machine breaks down. They will have to race home before the fish go off. Back on shore, a television weatherman has noticed something very odd happening. Hurricane Grace is about to collide with two other major weather systems. 'You could be a meteorologist all your life and never see something like this', he gasps. 'It would be a disaster of epic proportions. It would be the perfect storm.'

Gloucester, Massachusetts, from where the ill-fated fishing boat in *The Perfect Storm* set sail.

The Storm (2009), inspired by real events in the Netherlands in 1953.

Tyne receives a severe storm warning, and tells the crew that if they want to get their fish back while it is still saleable they will have to battle through 50-ft waves. The men decide unanimously to go for it. Tyne loves captaining a fishing boat. 'Is there anything better in the world?' he asks Linda, a fellow skipper, early in the movie. Now comes a long and genuinely scary storm sequence, mainly concentrated on the *Andrea Gail*, but also including the rescue of three people from a yacht and the loss of a crewman when a rescue helicopter has to ditch in the sea.

At first Tyne and his crew find the fight against the weather exhilarating, but as waves bounce the fishing boat around, smash in windows and tear equipment from the decks, the skipper finally announces they will have to give up their attempt to get home and reverse their course. But it is too late. They are confronted by an especially fearsome wave, towering over them like a mountain, with an ominous overhang at the top. Laboriously the *Andrea Gail* manages to climb about halfway up, but then she overturns, and everyone aboard is lost. At the memorial service in Gloucester, the congregation sings: 'Oh hear us when we cry to Thee for those in peril on the sea', and Linda delivers a moving, tearful address. The men who have died lie in a vast unmarked grave. They have no headstones where we can lay flowers. We can visit them only in our hearts and our dreams. Finally we see that their names have been added to the long list of those who have perished off Gloucester. The real storm at the heart of the film was the 'Halloween nor'easter' of 1991, which absorbed Hurricane Grace, driving winds to 160 km/h (100 mph) and waves to 30 m (100 ft). Nine people died, including the six-man crew of the *Andrea Gail*.

Another film based on a real event was *The Storm*, a Dutch production, almost documentary in style, made in 2009 about the storm surge of 1953 that killed 1,800 people in the Netherlands, as well as more than 600 in the UK and around its coasts. We do see teeming rain, howling winds, crashing waters, wrecked houses and dead bodies floating in the wreckage, but the central (fictional) story concerns a young, single mother named Julia and her baby, Ernest. As the waters rise at their farmhouse, drowning her mother and her sister, Julia tries to protect her baby by putting him in a little chest as high up as she can, then ties herself by a rope to a wooden grille on what is left of the collapsing building. But the dead body of a cow gets entangled with the grille, pulling it down and dragging her into the waters. She finally manages to break free and tries to swim back, but is unconscious and drowning when Aldo, a navy lieutenant, leaps down from a helicopter to save her. Julia comes round in a military hospital and is desperate to find Ernest. Aldo, who happens to be the brother of the baby's father, helps her as best he can, risking death on flooded causeways, in leaking boats, or wading chest deep through water. Eventually they find the chest, but there is no baby inside. Ernest has been taken by another woman whose own child had been killed a few weeks before in a car crash.

At the local hotel, which has become an emergency refuge, she hides the baby from Julia, who is treated with a distinct lack of compassion. Before the flood, she had been shunned as a 'whore', who has brought disgrace on, well, just about everybody. Now other young mothers mutter that the storm is God's punishment for her sins. Indeed, there is little glamorization of the survivors. The hotelier wants to charge full prices for the food he provides, and seems more interested in the business's bottom line than anything else. The 1953 sequence ends with Aldo and Julia being taken away from the hotel by boat with other survivors, though some do not want the young woman with them. The action then cuts to 1971, when new flood defences are being opened. Decorations are handed out to the heroes of 1953. Aldo gets one, as does the hotelier. Julia encounters the woman who took her baby all those years ago, and she breaks down and admits what she did. Mother and son meet again at last, and Julia is magnanimous in recognizing that the other woman has brought him up for eighteen years. The boy asks whether Aldo is his father. Julia says no, but that it was he, Ernest, who brought them together. Julia and Aldo seem happily married, but there is no sign that they have children. At the close of the film Ernest stands uneasily in the middle of a wide void between his two mothers.

Paintings and films may be all very well, but some people want to see tempests much more up close and personal, and the 1950s saw the birth of a new hobby: storm chasing. A U.S. Environmental Protection Agency official, David Hoadley, is often regarded as its father. He got hooked when he was seventeen, filming a huge thunderstorm rampaging through his home town of Bismarck, North Dakota, on an 8 mm camera. From that he graduated to hanging around the local weather bureau, looking over the shoulders of meteorologists, trying to work out when and where the next tornado would appear. Hoadley said that 'few life experiences' could compete with witnessing a fierce storm: 'the sheer, raw experience of confronting an elemental force of nature – uncontrolled and unpredictable'.[13] His newsletter *Storm Track* helped to popularize the pastime, which also got a leg-up from the 1996 feature film *Twister*, in which tornado chasing reunites a couple whose marriage is breaking up.

A storm chaser in action, Johnson City, Kansas, 24 May 2015.

Soon tour companies sprang up to ferry people to close encounters. The director of one claims he has seen more than 630 tornadoes. In 2007 the Discovery Channel started running the series *Storm Chasers*. But it was new technology that probably gave chasing its biggest boost. The Internet helped devotees to form communities, while mobile phones enabled them to direct each other to where a storm was brewing. Laptop computers provided access to fast-changing weather information in their

vehicles, GPS helped them get to the right place quickly, while smaller and cheaper cameras meant that more and more chasers could film and photograph storms. Footage of one tornado posted on YouTube attracted more than 3 million hits. Chasers can provide valuable information that helps weather forecasters deliver more accurate predictions, and some meteorologists believe that showing live footage of a developing storm may make viewers take warnings more seriously.

But storm chasing remains hazardous, even for experts. Reed Timmer, who specializes in driving armoured vehicles into tornadoes, says he has had his eardrums rupture because of the dramatic fall in pressure. And popularity has made some hazards worse. David Hoadley feels that chasing has 'turned into a bit of a circus. And with crowded, muddy roads where it's possible to get trapped in a storm, there's good reason to be concerned.' Tim Samaras, a storm-chasing author who had featured on Discovery's programmes, complained: 'On a big tornado day in Oklahoma, you can have hundreds of storm chasers lined up down the road.'[14] Samaras had contributed a lot to our knowledge of tornadoes by placing probes in their path. In 2003, on the outskirts of Manchester, South Dakota, one recorded a 100-millibar drop in pressure – the biggest ever captured at the time – as the little town was, as he put it, 'literally sucked into the clouds' by an F4 tornado.[15] With 25 years of chasing under his belt, he was working with his son Paul, aged 24, and his colleague Carl Young during the tornado season of 2013. Samaras was known for his caution, always urging others to be careful, and never hesitating to abort a mission if he felt it was too risky, sometimes to the irritation of colleagues. At about six in the evening of 31 May, the trio spotted a twister near El Reno, Oklahoma. A few days before in Kansas, they had seen a tornado that was a 'gorgeous, glowing tangerine against the sun while its long rope undulated like a belly dancer', but this one was less photogenic – 'a black wedge of indistinct composition'. It was wrapped in rain so the Samaras team could not see which way it was heading, nor that it was crushing buildings, uprooting telegraph poles and flinging vehicles around. The last audible comment on their camera came from Young: 'Wow. What a beast.' Their smashed car was found about an hour later. Tim Samaras was dead inside. The bodies of Paul and Carl were discovered in a ditch a quarter of a mile away. Another twenty people perished in the area.

7 Futures

'Being an island nation at the eastern edge of the Atlantic ocean, we're used to getting storms during autumn and winter, but not relentlessly for such long periods of time.'[1] So said *Channel 4 News* weatherman Liam Dutton of the UK winter of 2013–14, which sent floodwaters rising and records falling. It began with the St Jude's storm of 27 October, which saw four people killed, including a seventeen-year-old girl who was crushed when a tree fell on a caravan, and a man and woman who were trapped under rubble after an uprooted tree caused a gas explosion. More than 600,000 homes had their power cut off. Just over a month later, a major storm on 5 December produced the worst tidal surge along England's east coast since 1953, with the waters in some places actually getting higher than they did in the devastating flood of the Coronation year. Fortunately, improved flood defences and forecasting kept the death toll down to two. More wild weather would bring a miserable Christmas for many. Seventy thousand homes had their power cut and four people died, including one man who was killed in a car crash during a hailstorm, and another who drowned trying to rescue his dog from a swollen river. For Britain it was the stormiest December since 1969, while for Scotland it was the wettest on record.

Nor did the New Year bring any respite. On 3 January the village of Muchelney in the Somerset Levels was cut off as the Parrett and Tone rivers burst their banks, bringing the worst flooding for nearly a century. Five days later, Home Counties towns such as Staines, Chertsey and Weybridge found themselves under water. Still the rain fell, and when the Environment Secretary Owen Paterson visited Somerset, he was confronted by angry residents protesting at a failure to dredge rivers for twenty years, which they said had caused the crisis. January 2014 turned out to be the wettest ever in central southern and southeast England. Then February came in like a lion, with a huge storm demolishing the main railway line to the West Country at Dawlish in Devon. The Thames rose to its highest level in 60 years, and Worcester too found itself partly submerged as the Severn overflowed. On the Somerset Levels flood defences were now overwhelmed in the village of Moorland, and residents from three other villages were warned to leave their homes. As the waters rose, so did the political temperature. The Prime Minister had to take personal control of the crisis, while one of his colleagues blamed the failure to dredge rivers on the Environment Agency. The Agency's boss retorted that it was the government's fault for failing to provide the money, adding that his staff had done 'a valiant job trying to cope with unprecedented natural forces'.[2] A Somerset MP called him a 'coward' and threatened to 'stick his head down the loo'.[3] And the storms had still not finished. On 15 February an 85-year-old

Damage caused by the St Jude's storm, October 2013, North London.

A house lifted off its foundations, New Orleans, after Hurricane Katrina, 2005.

passenger was killed when a freak wave crashed through the window of a cruise ship in the Channel, and a taxi driver died in central London when her car was crushed by falling masonry.

So was this just an exceptional year, or are storms getting fiercer? Certainly in Britain four of the five wettest years since records began have occurred since 2000, and the rest of the world too has been experiencing wilder weather. Take the Americas, where in 2004, for the first time ever, a hurricane was spotted developing off the coast of Brazil. It made landfall about 800 km (500 miles) from Rio de Janeiro, killing three people and destroying crops as well as about 1,500 homes. The following year, an American meteorologist reported that storms in the north Atlantic and the western north Pacific were 50 per cent more powerful and were lasting 60 per cent longer than in 1949, though this might be partly because we were getting better at spotting those that formed at sea and never made landfall. In August 2005 Hurricane Katrina became the costliest storm in history. With wind speeds of more than 270 km/h (170 mph), it gouged great holes in New Orleans' flood defences, submerging about four-fifths of the city. More than 1,800 people died, and the damage was estimated at about $80 billion. According to some estimates, only two natural disasters of any kind have done more damage.

In 2010 up to 20 million people were caught up in floods in Pakistan caused by the most torrential monsoon rains in 80

Flooding in Pakistan, September 2010.

years. As about 1,750 people perished, and more than a million homes were destroyed, John Holmes, the United Nations Under-Secretary-General for Humanitarian Affairs, described the disaster as 'one of the most challenging that any country has faced in recent years'.[4] While Pakistan was still reeling, devastating monsoon rains struck again in the autumn of 2011, killing 400 people and destroying hundreds of thousands more homes. In January of that year more than 500 people had already died in mudslides that followed some of Brazil's fiercest rainstorms in decades, while months of downpours caused what Colombia's president, Juan Manuel Santos, called 'the worst natural disaster that we can remember'. He said 3 million people had been caught up in what felt like a hurricane 'that hasn't wanted to leave'.[5] Beginning in March 2011, some parts of Thailand had rains that were more than four times as severe as normal, and then in August came Tropical Storm Nock-Ten, generating the country's worst floods in half a century. The death toll was 800, with 13 million people caught up in the disaster. More than 550 were killed by tornadoes in the United States, the worst total since

1925. In August the country's east coast was pummelled by Hurricane Irene, the first natural disaster to shut down New York City's subway system. It also forced the evacuation of more than 2 million people, and became the sixth costliest hurricane in the country's history, while record-breaking October blizzards, 'Snowtober', would see 75 cm (2½ ft) of snowfall in some parts of the northeastern USA. The same month, torrential rains brought flash floods that drowned nine people in Italy. From June until September, China suffered its fiercest rainstorms in more than half a century, with Sichuan experiencing its worst flooding since records began in 1847. More than 350 people perished. A week before Christmas, one of the worst typhoons ever to hit Mindanao in the Philippines killed 1,000, and left 50,000 homeless.

All in all, 2011 was quite a year. Indeed, the global reinsurance company Munich Re said it was the world's costliest ever for natural disasters. According to Christopher Vaccaro of America's National Oceanic and Atmospheric Administration (NOAA), it 'rewrote the record books. From crippling snowstorms to the second deadliest tornado year on record to epic floods, drought

The Thailand floods of 2011, the worst in half a century.

and heat, and the third busiest hurricane season on record, we've witnessed the extreme of nearly every weather category.'[6] Then the following year saw the biggest storm ever in the Atlantic – Hurricane, or 'Superstorm', Sandy – which was 1,450 km (900 miles) across and became the costliest in history apart from Katrina. In November 2013, just as a United Nations summit on fighting global warming began, the Philippines were battered by Typhoon Haiyan. With gusts hitting nearly 320 km/h (200 mph), some considered it perhaps the strongest storm ever to make landfall. More than 4,000 people were killed, and about 600,000 were made homeless, while President Benigno Aquino declared a 'national calamity'.[7] The leader of the Philippines delegation at the climate talks, Naderev Sano, declared that 'the trend we now see is that more destructive storms will be the new norm', and he even went on hunger strike for two weeks to try to cajole the summit into making meaningful progress.[8]

His view is lent support by a study from the Environment Maryland Research and Policy Center, which found that the number of severe downpours over the continental United States had increased by nearly a quarter between 1948 and 2006. Worldwide the United Nations declared that while the frequency of Category One storms remained about constant from 1970 to 2004, the number in the strongest categories – Four and Five

East River Park, New York City, after Hurricane Irene, August 2011.

Blizzard in Pen Argyl, Pennsylvania, during 'Snowtober', October 2011.

– nearly doubled. In theory global warming should lead to fiercer storms, because the hotter it is, the more water evaporates into the air, and warmer air can hold more vapour, meaning that when downpours happen they are likely to be heavier. It does seem pretty clear that the world is warming up, though there is still argument over whether human beings are to blame. In 2011 the NOAA pointed out that the first eleven years of the twenty-first century were all among the hottest thirteen since records began in 1880, and two years later, data from the administration showed 2013 was the fourth-hottest year on record, and that every one of the last 37 had been warmer than the average for the twentieth century. In the UK the Chief Scientist at the Meteorological Office, Dame Julia Slingo, said that 'all the evidence' suggested that the UK winter storms of 2013–14 were linked to climate change, and Prime Minister David Cameron said that he too 'suspected' that was the case.[9]

Remains of the rollercoaster at Seaside Heights, New Jersey, after 'Superstorm' Sandy struck in October 2012.

Probably the most authoritative assessment on the subject comes from the United Nations Intergovernmental Panel on Climate Change, which seeks a consensus from the views of thousands of eminent scientists all over the world. For a body that has stirred up such fury among those who disagree with its findings, it often expresses itself with an almost soporific sobriety. In its 2012 report *Managing the Risks of Extreme Events and Disasters to Advance Climate Change Adaptation: Summary for Policymakers*, it murmurs: 'It is likely that the frequency of heavy precipitation or the proportion of total rainfall from heavy falls will increase in the 21st century over many areas.'[10] When it comes to the ferocity of the fiercest storms of all – tropical cyclones – the report says their wind speeds will probably increase, though their number may not. The NOAA's Geophysical Fluid Dynamics Laboratory made a similar prediction: although there might be no more of them, 'the strongest hurricanes in the present climate may be upstaged by even more intense hurricanes over the next century.'[11] But even if cyclones were to get no stronger than they are now, they would still be likely to become more devastating. That is because sea levels are rising, making storm surges more lethal, and because every day there are 200,000 more humans, bringing 'increasing exposure of people and economic assets'.[12] An analysis of what would happen if the Britain of today was struck by a tempest like the Great Storm of 1703 concluded that the scale of casualties and damage would be 'catastrophic', with 18 million properties in harm's way.[13] The IPCC report also makes it clear that another trend will continue: people in poor countries will suffer more than those in rich ones. One of the report's main editors said that storms, combined with rising seas, could make some areas uninhabitable. Mumbai was one of a number of cities named as particularly vulnerable. Stanford University climate scientist Chris Field warned that some islands might have to be abandoned: 'The decision about whether or not to move is achingly difficult', he said, 'and I think it's one that the world will have to face with increasing frequency in the future.'[14] Earlier IPCC reports had made more precise predictions, with one in 2007 suggesting that the intensity of cyclones over Asia would increase by 10 to 20 per cent. Scientists now had 'a better understanding of uncertainties', according to David Karoly at the University of Melbourne. But he does not see this as an excuse for inaction: 'It doesn't really matter if the car hits the wall at 70 or 80 kilometres an hour. You should still wear your seat belt.'[15]

Street in Genoa during the flash floods of November 2011.

But fighting climate change has proved very difficult. So what if we humans could find other ways of controlling storms? As we have seen, in ancient times we attempted to do it through gods. Then came more rational approaches. In the late nineteenth century they fired mortars in the vineyards and orchards of Central Europe to try to prevent hailstorms devastating crops because of a theory that shockwaves in the atmosphere could disrupt the formation of hailstones. Great success was claimed, until a series of scientifically supervised experiments led to the method being dismissed as useless. The Soviet Union took a more ambitious approach to protect the cotton fields of Uzbekistan, firing rockets and artillery shells into the clouds. These carried silver or lead iodide crystals and the idea was to provide many different nuclei around which hailstones would form. This would result in more, but smaller stones. Many would melt before they reached the ground, while those that did complete the journey would be so tiny they would do little damage. The Russians also tried this approach in other areas such as Georgia and Moldavia, and they too claimed spectacular success, saying that between 1968 and 1984 they had reduced the damage from hailstorms by 80 per cent, but tests in the USA were unable to reproduce the Soviet findings. In southwest France, they tried seeding clouds from generators on the ground, and an examination of insurance claims for 1965 showed they were 41 per cent lower in the areas that had been seeded than in others that had not. In Switzerland, on the other hand, seeding seemed to produce more hail. An unexpected hazard of this method was demonstrated in 2008 when the Russian air force tried using cement in sacks for seeding. One failed to open in the air and instead went crashing through the roof of a house.

Dust storms can be combated in a less high-tech way. Following the dreadful 'Black Blizzards' that blotted out the sun in America's Dust Bowl in the 1930s, the government invested heavily in planting windbreaks and restoring grasslands. And after China was plagued by dust storms in the 1950s, the authorities began planting a vast belt of forests, the 'Great Green Wall' across the arid lands in the north of the country. Great

Hail damage to a field of corn in Veldenz, Germany, 2011.

walls – three of them – were also proposed in America to tame tornadoes. Professor Rongjia Tao of Temple University, Philadelphia, suggested they needed to be nearly 300 metres (1,000 ft) high and up to 160 kilometres (100 miles) long, with one to be built in North Dakota, one along the Kansas–Oklahoma border and a third in Texas and Louisiana. The idea would not be to shield towns from twisters. The walls would not be strong enough. Instead they would be designed to stop tornadoes forming by disrupting the clash of hot southern and cold northern air. The walls would cost nearly $10 billion to build, but the professor said they would soon pay for themselves because of the damage they would prevent. His proposal was floated at the American Physical Society meeting in Denver in 2014, and quickly rejected by other scientists as theoretically unsound as well as impracticable.

The jewel in the crown of storm control would be the taming of hurricanes. In the 1940s the Americans began experiments that involved seeding the clouds around the eye of a hurricane to make it bigger, and so slow down the winds whirling around it. But interfering with tropical storms is a high-risk strategy.

Countries such as Japan and Mexico depend on them for a lot of their rainfall, and the U.S. also found problems closer to home. Following allegations that the seeding of a hurricane had made it change course and devastate Georgia in 1947, a legal battle broke out in December 1955 when fierce storms brought widespread flooding to northern and central areas of California, with 64 people killed, and damage estimated at $200 million. Some of those affected tried to sue the state and a number of organizations that had been involved in cloud seeding at the time.

Nothing daunted, in 1962 the U.S. government launched Project Stormfury to test hurricane-seeding systematically. Reductions of up to 30 per cent in wind speeds were observed on some days, though at other times there was no change, and sometimes the storms even seemed to get stronger. Critics said that when a storm did weaken there was no evidence it was caused by seeding, and in 1983 the project was cancelled. A variety of other ideas have been suggested since, such as spreading soot across clouds to absorb sunlight, or building huge wind farms off coasts to steal energy from hurricanes and make them less destructive. Alternatively perhaps pointing jet engines up into the clouds to trigger smaller storms would disrupt a big storm's progress. Or what about coating the ocean with an oily film to prevent evaporation, or pouring liquid nitrogen onto it, or even pumping up colder water from the depths to cool it down and deprive the hurricane of the heat that keeps it alive? At present there is no great confidence that any of these ideas would be effective or even practicable.

The IPCC's suggestions for limiting the destructive power of extreme storms are much more prosaic – like choosing more sensibly where we build, making the structures we erect more robust and improving drainage. Other ideas include better forecasting and more effective dissemination of warnings, conserving natural protection like mangroves, or simply constructing better roads so people can get out of harm's way quicker. It all sounds much less space-age than shooting missiles into the clouds or pumping up water from the deep ocean, and perhaps one day such things may be shown to work, but as Frank Marks, director of NOAA's Hurricane Research Division, warned: 'I don't think we know enough about the natural balance, even after studying it for 50 years at NOAA.'[16] He believes that for the moment at least, weakening great tropical storms is beyond our power: 'I have more respect for how complicated the process is. It's pretty humbling, considering how little we know.'

NOTABLE STORMS

c. 524 BC
Sandstorm buries army of Cambyses II, Egypt

AD 394, 6 September
Dust storm at Battle of the Frigidus River on the border of Italy and Slovenia

c. 400
Niya buried by sandstorm, China

876
Storm wrecks 120 Viking ships, Swanage, UK

1274, 1281
Storms wreck Mongol invasion fleets, Japan

29 March 1461
Blizzard at the Battle of Towton, UK

1502
25 Spanish ships sunk in hurricane in the Americas

26 November 1703
Great Storm, UK

10 October 1780
Great Hurricane, the Caribbean

6 November 1856
Lightning strikes gunpowder store, Rhodes

12 January 1888
Great Schoolhouse Blizzard, USA

30 April 1888
Moradabad hailstorm, India

18 March 1925,
Great Tri-State Tornado, USA

31 January 1953
North Sea Storm surge, England and Netherlands

12 November 1970
Bhola Cyclone, Bangladesh

24 December 1971
Lightning brings down airliner, Peru

February 1972
Ardekan blizzard, Iran

3 April 1974
148 tornadoes hit USA

16 October 1987
Great Storm, UK

26 April 1989
Saturia–Manikganj Sadar tornado, Bangladesh

29 April 1991
Bangladesh cyclone

August 2005
Hurricane Katrina

2 May 2008
Cyclone Nargis, Myanmar

August 2011
Hurricane Irene

October 2012
'Superstorm' Sandy

REFERENCES

1 Religion

1 Stu Ostro, 'Weather Phobias', www.weather.com/blog, 7 July 2006.
2 Rosalind Fergusson, *The Penguin Dictionary of Proverbs* (Harmondsworth, 1983), p. 255.
3 Judika Illes, *The Encyclopaedia of Spirits* (New York, 2009), p. 294.
4 *The Epic of Gilgamesh*, trans. N. K. Sandars (Harmondsworth, 1975), pp. 108–12.
5 Sir Ernest A. Wallis Budge, *The Babylonian Legends of Creation* (New York, 2010), pp. 51, 52.
6 Simon B. Parker, 'Syrian and Palestinian Religion', *Encyclopaedia Britannica*, http://library.eb.co.uk.
7 Sir James George Frazer, *The Golden Bough: A Study in Magic and Religion* (London, 1963), p. 212.
8 Ibid.
9 Kerry Emanuel, *Divine Wind: The History and Science of Hurricanes* (Oxford, 2005), p. 30.
10 Matthew 8:20–24; Mark 4:38; Matthew 8:26; Mark 4.39; Acts 27:14–22.
11 Iona Opie and Moira Tatem, *A Dictionary of Superstitions* (Oxford, 1996), www.oxfordreference.com.
12 Jacqueline Simpson and Steve Roud, 'black dogs', in *A Dictionary of English Folklore* (Oxford, 2003), www.oxfordreference.com.
13 Frazer, *Golden Bough*, p. 867.
14 Pliny the Elder, *The Natural History*, trans. John Bostock and H. T. Riley (1855), Book 28, ch. 23, www.perseus.tufts.edu.
15 Frazer, *Golden Bough*, p. 107.

2 Nature

1 Aristotle, *Meteorology*, trans. anon. (Whitefish, MT, 2004), pp. 21, 31.
2 John D. Cox, *Storm Watchers: The Turbulent History of Weather*

Prediction from Franklin's Kite to El Nino (Hoboken, NJ, 2002), p. 31.
3 John Gribbin, *Weather Force* (London, 1979), p. 77.
4 John Timbs, ed., *Arcana Science Art, Or Annual Register Popular Inventions Improvements*, vol. v (Charleston, SC, 2011), pp. 253, 252.
5 Gribbin, *Weather Force*, p. 80.
6 Ibid., p. 34.
7 Angela Balakrishnan, 'China's Stock Market Battered by Storms', *The Guardian*, 1 February 2008, www.guardian.co.uk.
8 'Caught on Video: The Moment a Driver is Swallowed by a Menacing Black Cloud as a Huge Dust Storm from the Desert Engulfs Phoenix', *Daily Mail*, 7 July 2011, www.dailymail.co.uk.
9 John Farrand Jr, *Weather* (New York, 1990), p. 189.
10 'Beijing Sandstorm 2010', *Huffington Post*, 19 May 2010, www.huffingtonpost.com.
11 Gribbin, *Weather Force*, p. 76.
12 Samuel Taylor Coleridge, *The Rime of the Ancient Mariner*, in *The Oxford Book of English Verse, 1250–1918*, ed. Sir Arthur Quiller-Couch (Oxford, 1961), p. 649.
13 *Hurricane – The Anatomy: Predicting the Unpredictable* (Saint Thomas Productions, 2014).
14 Ibid.
15 Michielle Beck, 'A Real-life Weather Storm Story', http://weather.about.com, accessed 31 July 2013.
16 Alice Jackson, 'One Survivor's Story', www.people.com, 8 September 2005.
17 John E. Oliver, ed., *The Encyclopedia of World Climatology* (Berlin, 2005), p. 156.
18 Saffir-Simpson Hurricane Wind Scale, www.nhc.noaa.gov.
19 U.S. Dept of Agriculture, Soil Survey, 1967, p. 183.
20 Jack Healy, 'Isaac Brings Touch of Relief, and Hope for Next Season, to Corn Belt', *New York Times*, 2 September 2012, www.nytimes.com.
21 Jack Watkins, '1987 Great Storm: Terrible Blow, Not a Knockout', *Daily Telegraph*, 13 October 2007, www.telegraph.co.uk.
22 Ibid.
23 'Hurricanes Bring Benefits to Barrier Islands and Beaches, Even as They Devastate Communities Located There, Duke Researchers Say', *Duke Today*, 13 September 2004, today.duke.edu.

3 Effects

1 Cicero's description. 'Translating Herodotus', *The Economist*, 21 September 2013.
2 Herodotus, 'Xerxes Invades Greece' from *The Histories*, trans. anon., www.fordham.edu.

3 Paterculus, *Roman History*, Book 2, chapters 117–20, trans. F. W. Shipley, www.livius.org.
4 Cassius Dio, *Roman History*, book 56, trans. Earnest Cary, www.livius.org.
5 Ibid.
6 Ibid.
7 Ibid.
8 'Publius Quinctilius Varus', *Encyclopaedia Britannica*, http://library.eb.co.uk.
9 Publius Annius Florus, *Epitome of Titus Livy*, www.livius.org.
10 Edward Gibbon, *The History of the Decline and Fall of the Roman Empire* (London, 1904), vol. III, pp. 214–17.
11 Ibid.
12 Ibid.
13 Ibid.
14 *The Anglo-Saxon Chronicle*, trans. G. N. Garmonsway (London, 1972), p. 75.
15 Jonathan Bardon, *A History of Ulster* (Belfast, 1992), p. 25.
16 Harold A. Winters, *Battling the Elements: Weather and Terrain in the Conduct of War* (Baltimore, MD, 2001), p. 15.
17 English Heritage Battlefield Report: Towton 1461, www.english-heritage.org.uk.
18 Ibid.
19 Garrett Mattingley, *The Defeat of the Spanish Armada* (London, 1983), p. 438.
20 Bardon, *Ulster*, p. 90.
21 William Strachey, *A True Reportory of the Wracke, and Redemption of Sir Thomas Gates Knight*, encyclopediavirginia.org.

4 Events

1 Nigel Cawthorne, *100 Disasters that Shook the World* (London, 2005), p. 64.
2 John Gribbin, *Weather Force* (London, 1979), p. 60.
3 Jonathan Head, 'Rebuilding Burma's Cyclone-hit Irrawaddy Delta', www.bbc.co.uk/news, 21 December 2012.
4 'Eyewitness: Burma Delta Disaster', www.bbc.co.uk/news, 20 May 2008.
5 Eric Schmitt, 'Gates Accuses Myanmar of "Criminal Neglect"', *New York Times*, 7 June 2008, www.nytimes.com.
6 Natalia Antelava, 'Burma Generals Failing Their People', www.bbc.co.uk/news, 15 May 2008.
7 'Monks Succeed in Cyclone Relief as Junta Falters', *New York Times*, 31 May 2008, www.nytimes.com.

8 Head, 'Rebuilding'.
9 Gavin Shorto, 'The Great Hurricane', *The Bermudian* (Fall 2010), www.thebermudian.com.
10 NEMO Secretariat, 'Saint Lucia: NEMO Remembers the Great Hurricane of 1780', 7 October 2005, www.cdera.org.
11 Shorto, 'Great Hurricane'.
12 'Survivors in Bangladesh Welcome Aid After Tornado', *Washington Post*, 2 May 1989.
13 Gribbin, *Weather Force*, p. 75.
14 Ibid., p. 80.
15 'India', *The Times*, 7 May 1888, p. 7.
16 Hari Menon, 'Bones of a Riddle', www.outlookindia.com, 8 November 2004.
17 'Are These the Bones of a Legendary Persian Army Lost in the Sahara 2,500 Years Ago?', *Daily Mail*, 10 November 2009, www.dailymail.co.uk
18 Charles Tomlinson, *The Thunder-storm: An Account of the Properties of Lightning and of Atmospheric Electricity in Various Parts of the World* (London, 1859), p. 168.
19 S.R.S., 'Life at Smyrna', *The Times*, 27 November 1856, p. 5.
20 Frederik Pleitgen, 'Survivor Still Haunted by 1971 Air Crash', CNN, 2 July 2009, www.edition.cnn.com.
21 Juliane Koepcke, 'How I Survived a Plane Crash', www.bbc.co.uk/news, 24 March 2012.
22 Pleitgen, 'Survivor Still Haunted'.
23 Koepcke, 'How I Survived'.
24 Martin Brayne, *The Greatest Storm* (Stroud, 2002), p. 4.
25 Kaari Ward, ed., *Great Disasters: Dramatic True Stories of Nature's Awesome Powers* (New York, 1989), p. 93.
26 Brayne, *Greatest Storm*, p. 73.
27 Oliver Duff, 'Novice Divers Find Wreck of 17th Century Warship', *The Independent*, 29 May 2006, www.independent.co.uk.
28 Gribbin, *Weather Force*, p. 66.
29 Brayne, *Greatest Storm*, pp. 8, 54, 118, 180, 187.
30 Ibid., pp. 180, 187.
31 Jo Macfarlane, 'Killer Storms to Lash Britain', *The Express*, 17 October 2007, www.express.co.uk.
32 Brayne, *Greatest Storm*, p. 129.
33 Ibid.
34 Ibid., p. 190.
35 Ibid., p. 104.
36 Ibid., p. 149.
37 Ibid., pp. 168, 50.

5 Literature

1 Daniel Defoe, *The Life and Adventures of Robinson Crusoe*, ch. 1, www.gutenberg.org.
2 Ibid., ch. 3.
3 Richard Hughes, *A High Wind in Jamaica* (London, 1992), pp. 4, 19, 22, 25.
4 Carol Birch, *Jamrach's Menagerie* (Edinburgh, 2011), pp. 196–200.
5 Guy de Maupassant, 'The Drunkard', http://classiclit.about.com.
6 Alexander Pushkin, 'The Snow Storm', www.cadytech.com.
7 Emily Bronte, *Wuthering Heights*, ch. 9, www.gutenberg.org.
8 Charlotte Bronte, *Jane Eyre*, ch. 23, www.gutenberg.org.
9 Kate Chopin, *The Awakening and Selected Stories* (New York, 1976), pp. 267–72.
10 Emily Dickinson, 'A Thunderstorm', www.online-literature.com.
11 Gerard Manley Hopkins, *Selected Poems* (London, 1975), p. 3.
12 A. S. Byatt, *Possession* (London, 1990), pp. 492–507.
13 Joesph Conrad, *Typhoon* (London, 1968), pp. 145, 163.
14 Ibid., p. 153.
15 Ibid., p. 150.
16 Ibid., pp. 165–203.

6 Spectacle

1 Lord Byron, 'Beppo: A Venetian Story', from *Lord Byron: Selected Poetry* (Oxford, 1994), p. 96.
2 Jon D. Mikalson, *Ancient Greek Religion* (Hoboken, NJ, 2011), p. 15.
3 *Tempests and Romantic Visionaries: Images of Storms in European and American Art* (Tucson, OK, 2006), p. 82.
4 Oskar Bätschmann, *Nicolas Poussin: Dialectics of Painting* (London, 1994), p. 95.
5 'Turner and the Sea', National Maritime Museum, London, 22 November 2013–21 April 2014, exhibition notes.
6 *Tempests and Romantic Visionaries*, p. 8.
7 Ibid., p. 66.
8 'Impressionism and Open Air Painting from Corot to Van Gogh', Museo Thyssen-Bornemisza, Madrid, 5 February–12 May 2013, wall note.
9 Steven Z. Levine, *Monet, Narcissus, and Self-reflection: The Modernist Myth of the Self* (Chicago, IL, 1994), pp. 64, 68.
10 Frank S. Nugent, 'Samuel Goldwyn Turns Nordhoff-Hall "Hurricane" Loose Across the Screen of the Astor', *New York Times*, 10 November 1937, www.nytimes.com.
11 Roger Ebert, '*The Day After Tomorrow*', *Chicago Sun Times*, 28 May 2004.

12 Anthony Lane, 'The Film File: *The Day After Tomorrow*', *New Yorker*, 6 July 2004.
13 Stefan Bechtel with Tim Samaras, *Tornado Hunter: Getting Inside the Most Violent Storms on Earth* (Washington, DC, 2009).
14 Melody Kramer, 'Are Storm Chasers "Crossing the Line?"', National Geographic News, 4 June 2013, http://news.nationalgeographic.co.uk.
15 Robert Draper, 'The Last Chase', *National Geographic*, November 2013, http://ngm.nationalgeographic.com.

7 Futures

1 'Waterworld: Britain's Wettest Winter on Record', www.channel4.com/news, 15 February 2014.
2 'Thames Flooding Fears: Thousands of Homes At Risk', www.channel4.com/news, 10 February 2014.
3 'Waterworld'.
4 'UN Launches $459m Pakistan Flood Appeal', www.bbc.co.uk/news, 3 March 2010.
5 'President: Colombia Hit by "Worst Natural Disaster"', www.nbcnews.com, 26 April 2011.
6 John Vidal, 'Environment World Review of the Year: "2011 Rewrote the Record Books"', *The Guardian*, 22 December 2011, www.theguardian.com.
7 'Worse than Hell', *The Economist*, 16 November 2013.
8 'The New Normal?', *The Economist*, 16 November 2013.
9 'Met Office: Evidence "Suggests Climate Change Link to Storms"' www.bbc.co.uk/news, 9 February 2014.
10 'Managing the Risks of Extreme Events and Disasters to Advance Climate Change Adaptation: Summary for Policymakers', Intergovernmental Panel on Climate Change, www.ipcc-wg2.gov/SREX/report, pp. 7, 11.
11 'Global Warming and Hurricanes: An Overview of Current Research Results', Geophysical Fluid Dynamics Laboratory, www.gfdl.noaa.gov.
12 'Managing the Risks'.
13 Robert Doe, *Extreme Floods* (Stroud, 2006), p. 201.
14 'Climate Change Panel Warns of Severe Storms, Heatwaves and Floods', *The Guardian*, 28 March 2014, www.theguardian.com.
15 Michael Slezak, 'World Must Adapt to Unknown Climate Future, Says IPCC', *New Scientist*, 13 April 2014, www.newscientist.com.
16 Robert Krier, 'Can Hurricanes Be Tamed? Scientists Propose Novel Cloud-seeding Method', www.insideclimatenews.org, 24 October 2012.

SELECT BIBLIOGRAPHY

Allen, Daniel, *The Nature Magpie: A Cornucopia of Facts, Anecdotes, Folklore and Literature from the Natural World* (London, 2013)

Bechtel, Stefan, with Tim Samaras, *Tornado Hunter: Getting Inside the Most Violent Storms on Earth* (Des Moines, IA, 2009)

Brayne, Martin, *The Greatest Storm* (Stroud, 2002)

Brontë, Emily, *Wuthering Heights* [1847] (London, 1966)

Bulfinch, Thomas, *Myths of Greece and Rome* (Harmondsworth, 1981)

Calvocoressi, Peter, *Who's Who in the Bible* (London, 1988)

Cawthorne, Nigel, *100 Disasters that Shook the World* (London, 2005)

Chopin, Kate, *The Awakening and Selected Stories of Kate Chopin* (New York, 1976)

Conrad, Joseph, *The Nigger of the 'Narcissus', Typhoon, 'Twixt Land and Sea* [1897] (London, 1968)

Defoe, Daniel, *Robinson Crusoe* [1719], www.gutenberg.org

Doe, Robert, *Extreme Floods: A History in a Changing Climate* (London, 2006)

Emanuel, Kerry, *Divine Wind: The History and Science of Hurricanes* (New York, 2005)

The Epic of Gilgamesh, trans. N. K. Sanders (Harmondsworth, 1975)

Farrand, John Jr, *Weather* (New York, 1990)

Frazer, Sir James, *The Golden Bough* [1890] (London, 1963)

Garmonsway, G. N., ed., *The Anglo-Saxon Chronicle* (London, 1975)

George, Hardy S., ed., *Tempests and Romantic Visionaries: Images of Storms in European and American Art* (Oklahoma City, OK, 2006)

Gribbin, John, *Weather Force* (London, 1979)

Holford, Ingrid, *British Weather Disasters* (Newton Abbot, 1976)

Hughes, Richard, *A High Wind in Jamaica* [1929] (London, 1992)

Lester, Reginald, *The Observer's Book of Weather* (London, 1958)

Mattingly, Garrett, *The Defeat of the Spanish Armada* (London, 1983)

Milne, Anthony, *London's Drowning* (London, 1982)

Reynolds, Ross, *Weather* (London, 2008)
Ward, Kaari, ed., *Great Disasters* (Pleasantville, NY, 1989)
Withington, John, *A Disastrous History of the World* (London, 2008)

ASSOCIATIONS AND WEBSITES

Department of Earth Sciences, UCL
www.ucl.ac.uk/earth-sciences

Dust Storms Group, School of Earth and Environment, University of Leeds
www.see.leeds.ac.uk

Exeter Storm Risk Group, University of Exeter
http://emps.exeter.ac.uk

Hurricane Project, Dept of Earth Sciences, Durham University
www.dur.ac.uk

Lightning Protection Research and Testing, Cardiff University
http://lightning.engineering.cf.ac.uk

Lightning Research Laboratory, University of Florida
www.lightning.ece.ufl.edu

Met Office
www.metoffice.gov.uk

National Centre for Atmospheric Science
www.ncas.ac.uk

National Severe Storms Laboratory
www.nssl.noaa.gov

National Wind Institute, Texas Tech University
www.depts.ttu.edu/nwi

NOAA Hurricane Research Division
www.aoml.noaa.gov/hrd

Sand and Dust Storm Warning Advisory and Assessment System Regional Center for Northern Africa, Middle East, Europe
www.bsc.es/earth-sciences/sds-was

Taiwan Typhoon and Flood Research Institute
www.ttfri.narl.org.tw

Tornado and Storm Research Organisation
www.torro.org.uk

Typhoon Research Department, Meteorological Research Institute, Japan
www.mri-jma.go.jp

University Corporation for Atmospheric Research
www2.ucar.edu

PHOTO ACKNOWLEDGEMENTS

The author and publishers wish to express their thanks to the below sources of illustrative material and/or permission to reproduce it.

Bigstock: pp. 48 (BCFC), 120 (Warren Price); © The Trustees of the British Museum, London: pp. 19, 77, 109, 116; Anne Clements Photography: p. 160; Department of Foreign Affairs and Trade, Australian Government: p. 162; Federal Emergency Management Agency (FEMA): pp. 57 (Andrea Booher), 63 bottom (Jocelyn Austino); Tony Hisgett: p. 26 (top left); Infrogmation: pp. 63 (top), 161; International Dunhuang Project: p. 97; iStock: pp. 6 (LICreate), 10 (Alexander Dunkel), 27, 70 (ZU_09), 166 (JanaShea); Jaontiveros: p. 24; Lili Jinaraj: p. 46; Paul Keleher: p. 153; National Archives and Records Administration (NARA), Washington, DC: p. 81; Neryl Lewis, RRT: p. 62; Library of Congress, Washington, DC: pp. 56, 73, 113; Louen: p. 99; Ministry of Defence: pp. 54–5 (Sergeant Dan Bardsley); NASA: pp. 60–61; National Archives: p. 50 (Franklin D. Roosevelt Library); National Oceanic and Atmospheric Administration (NOAA) National Weather Service (NWS) Collection: pp. 49, 95; National Oceanic and Atmospheric Administration (NOAA) Photo Library: pp. 8 (Harald Richter), 52 (Sean Waugh NOAA/NSSL); PvdV: p. 115; Reinhardhuake: p. 29; Ferdinand Reus: p. 53; REX Shutterstock: pp. 51 (Everett), 86 (ZUMA), 100 (Everett), 156–7 (ZUMA), 167 (Davide); Matt Saal: pp. 40–41; Hiroshi Sanjuro: pp. 42–3; Christian Schröder: p. 34; Daniel Schwen: p. 25; David Shankbone: p. 164; State Library of Victoria: p. 126 (Allan C. Green collection of glass negatives); Nicholas A. Tonelli: p. 165; United States Department of Defense: pp. 83 (PHI Doolittle, USN), 163 (Cpl Robert J. Maurer); Victoria & Albert Museum, London: p. 108; United States Marine Corps: p. 45; Th. Walther: p. 58.

INDEX

Adad 15
Aeneas 12, 119
Aeolus 31
Afghanistan *54–5*, 96
air crashes 44–5, 100–101
Alaska 30, 32
Alba Longa 15
American War of Independence 92–3
Anaxagoras 35, 43
Arbogast 72–4
Aristotle 35
Atlantic Ocean 38, 58–9, 83, 161
Australia 45–6, 53
Aztecs 22

Baal 15, *15*
Bahamas *144*
Bakhuizen, Ludolf *134*, 135
Bangladesh 44, 61, *86*, 87–9, 93–4
    Concert for Bangladesh 89
Bantu 31
Barbados 91–2, 111
Beaufort Scale 65
Bedouin 32
Bermuda *83*, 83–5, 110
Berry, James 121–2
Bhola cyclone 87–9
Bible, the 26–9
Birch, Carol 116–18
Bjerknes, Vilhelm and Jacob 36–7
Bogor 42, *43*
Bohemia 33
Boreas 12, *13*
Borneo 32
Brazil 45, 161, 162
Brescia 99–100
Brittany 140
Brontë, Charlotte 119
Brontë, Emily 119
Bryant, William Cullen 122
Byatt, A. S. 122–3

Cairngorm Summit 65
California 24, *40–41*, 51, 170
Cambyses II, 'lost army' of 98
Canada 31, 38, 44, 49, 130
Capper, Col. James 36, 38
Caribbean, the 21, 64, 91–3
Carthage 12
Chac 24
Chenzhou 48
Chichen Itza 24, *25*
China 20–21, 46, 48, 52, 163, 168
Chopin, Kate 119–21
Churchill, Sir Winston 9
clouds 39, *40–41*, 42–4
cloud seeding 168–70
Coen Brothers (Joel and Ethan) 148–9
Cole, Thomas 138–9, *140*
Coleridge, Samuel Taylor 58, 121

Colombia 162
Columbus, Christopher 21, *22*
Conrad, Joseph 124–7
Constable, John 138, *138*
Cornwall 58, 102
crops, damage to 44, 50, 168
Cropsey, Jasper 138, *139*
Curry, John Steuart 143
Cyclone Nargis 61, *62*, 89–91
cyclones *see* hurricanes

Dawlish 160
Day After Tomorrow, The 147–8, *148*
Defoe, Daniel 104–5, 111–13
Dickinson, Emily 121
drought 15, 50
Dufy, Raoul 142
dust storms *48*, *50*, 50–51, 82, 168

Eddystone Lighthouse *101*, 102
Egypt 20, 52, 98
Elijah 18
Enlil 13–14
Espy, James Pollard 36
explosions caused by lightning
    99–100

Finland 18, 31
Florida 57, 60, 62, 152–3
Ford, John 144, 146–7
forecasting 80–81
France 32, 33
Frazier, Sir James George 32
French Revolution 82–3
Frigidus River, Battle of the 72–4
Fujita Scale 65

Galilee, Sea of 28, 124, 130
Germany 17, 31, 33, *34*, 52, 70–72
Gilgamesh, Epic of 14, 26
Giorgione *128*, 129–30
Grant, Sir Robert 7
Great Hurricane of 1780 91–3, *94*
Great Storm of 1987 66, 102
Great Storm of 1703 102–6, 167
Great Tri-State Tornado 94–6, *95*
Greece, ancient 11–12, 15, 31, 35,
    69–70, 130
Greenland 32

hail 26, 35, *37*, 43–5, 82, 96–7, *148*
Haiti 46
Halley, Edmond 36, 38
Halloween Nor'easter of 1991 155
Hassam, Childe 144, *146*
Hawes, Ray 66
Heade, Martin Johnson 139, *141*
'*Hesperus*, The Wreck of the'
    (poem) 123–4
High Wind in Jamaica, A 113–16
Hittites, the 12, 15
Hoadley, David 156, 158
Homer, Winslow 144, *145*
Hopkins, Gerard Manley 122
Hughes, Richard 113–16
Huracan 21–2
Hurrians, the 12
Hurricane, The (film) 144, 146–7,
    147
hurricanes 21–2, 58–67, 75–6, *144*,
    150–52
    Hurricane Andrew 60
    Hurricane Catarina *60*
    Hurricane Gilbert *114–15*
    Hurricane Grace 155
    Hurricane Irene 66, 163, *164*
    Hurricane Isaac 66
    Hurricane Ivan 62, *63*
    Hurricane Katrina 60, 61,
    62–4, *63*, *161*, 161, 164
hurricanes, naming of 64
Hyderabad 44

ice storms 49
Idrimi, king 15
Illinois 94–6
India 19–20, 36, 44, *46*
Indian Ocean 59

Indra 19
Iona (Inner Hebrides) 75
Iran 12, 69–70, 96
Iran hostage crisis, first rescue attempt 81–2
Iraq 12–14
Ireland 79
Italy *132*, 163, *167*

Jamaica 113–16, *114–15*
Jamrach's Menagerie 116–18
Jane Eyre 119
Japan 20, 66, 75–6, 170
Java 42
Jerome, Jerome K. 7
Jonah 27–8, *28*, *29*
Julius Caesar 107–9
Jupiter 12, 15, 17
Jurakan *see* Huracan

Kansas 55, 143, 150, *156–7*
Kelley, Dr Owen 59, 61
Kerala *46*
Key Largo 152–3, *152*
Khoikhoi people 32
Kiev 18
King Lear *109*, 109–10
Klimt, Gustav 142, *144*
Koepcke, Juliane 100–101
Korea, South 46
Krishna 20
Kublai Khan 75–6, *77*

Lei Gong 20–21
Leutze, Emanuel 139, *141*
Life of Pi 150–51, *151*
Lithuania 18
London 103–5
Longfellow, Henry Wadsworth 123–4
Louisiana 61, 119–21, *120*

Mabey, Richard 66
Macbeth *107*, 108
Malaysia 46
Malta 29
Maori 25–6, 30
Marduk 14
Marks, Frank 170–71
Massachusetts 57, 153–5, *153*
Maupassant, Guy de 118
Maya 24
Mexico 22, 24, 148, 170
Monet, Claude 140, *142*
Montana 47
Moradabad 96
Morrison, Jim 8
Moses *27*
Mozambique 31
Mumbai 167
Munch, Edvard 140, 142, *143*
music 8
Myanmar 61, *62*, 89–91

Nargis, cyclone 61, *62*, 89–91
National Oceanic and Atmospheric Administration (NOAA) 163, 165, 166, 171
Navajo 53
Nebraska 47
Netherlands 155–6
New Jersey 66, *166*
New Orleans 161
New Zealand 25–6, 30
New York *49*, 144, *146*, 148, 163, *164*
Nigeria 25
Ninurta 13, 15
Niya *97*, 98–9
Noah 26, 135
Nootka people 31
Normandy 118
Northampton 103
Northumberland 39
Novgorod 17

oak *10*, 11, 17, 18, 33
Oklahoma 54, *57*, 158
Oreithyia 12, *13*

Pacific Ocean 26, 59, 81, 150, 161
Pakistan 46, 161–2, *162*
Paraguay 32
Paul, St 28–9
Perfect Storm, The (film) 153–5
Perkunas (Perkuns) 18
Persia *see* Iran
Peru 100–101
Perun 17–18
Philippines 58, 163, 164
Phoenix *48*
Pilkey, Orrin 66–7
Possession 122–3
Poussin, Nicolas 133, *134*, 135
Preston 45
Project Stormfury 170
Pushkin, Alexander 118–19

rain 45–6, 161–2
Rainier, Mount 47
Rankin, Lt-Col. William 43
Redfield, William 36, 38
Rembrandt 130, *131*
Réunion 45
Rhodes 100
Robinson Crusoe 111–13, *113*
Rolling Stones, The 8
Rome, ancient 12, 30, 70–74
Rousseau, Henri 142–3, *145*
Rubens, Peter Paul *13*
Russia 118–19, 168

Saffir-Simpson scale 65
Sahel, the 51, *53*
St Jude's storm 159, *160*
St Lucia 92
Salmoneus, king 15
Samaras, Tim 158
sandstorms 51–2, *54–5*, *97*, 98–9
'Schoolhouse Blizzard' 47–8
Scotland 30, 42, 52, 79, 148
Seathwaite (Lake District) 45
Second World War 64, 76, 80–81
Serious Man, A 148–9, *149*
Seth 20
Shakespeare, William 85, 107–10
Shango 25, *26*
Shetland 31
Silver Lake, Colorado 47
Siwa 98, *99*
Slingo, Dame Julia 165
snow 46–8, *49*, 65, 96, *146*
'Snowtober' 48, 66, 163, *165*
Somerset 103, 159–60
South Africa 46
Spanish Armada 78–80, *80*
sport 8
Sri Lanka 46
Steinbeck, John 51
Storm, The (film) 155–6
storm chasers 156–8, *156–7*
Storm on the Sea of Galilee 130, *131*
storm surge, 1953 (Netherlands) 155–6, 159
Stravinsky, Igor 18
Suffolk 30
Sumatra 32
superstitions, storm 29–33
'Superstorm' Sandy 66, 164, *166*
Susano-o-no-Mikoto 20
Swanage *73*

Taino people 21–2
Taranis 18
Tarhun 12, 15
Tāwhirimātea 25–6
Tempest, The 85, 110, *112*
Teshub 12
Teutoburg Forest, Battle of the *68*, 70–72
Texas 57
Thailand *45*, 46, 162, *163*
Thales of Miletus 35
Theodosius I 72–4
thunderbird, the 24, *26*
Thor *16*, 17
Thrym 17
Tiamat 14

Tlaloc 22, *24*
Tlingit people 30
tornadoes *8*, *52*, 53–7, *56*, *57*, 143, 149–50, 162–3
  Great Tri-State Tornado 94–6, *95*
  Saturia-Manikganj Sadar tornado 93–4
Towton, battle of 76–8, *78*
tropical cyclones *see* hurricanes
Troy 12
Tsonga people 31
Turkey 12, 15
Turner, J.M.W. 135–8, 140
Typhoon 124–7
Typhoon Haiyan 164
typhoons *see* hurricanes
Tyrol, the 31

Uganda 46
Ukko 18
Ulysses 31
United Kingdom 18, 38–9, 44, 57, 159–61, 165
United Nations Intergovernmental Panel on Climate Change (IPCC) 166, 170
United States 38, 44–5, 48, 49–51, 56, 61

Vancouver 47
Varus, Publius Quinctilius 70–72
Velde, Willem van de, the Younger *23*
Vernet, Claude-Joseph 135, *136*, 138
Vikings 75
Vladimir I (the Great) 18
votive pictures 130, *132*

Wales 102
war 8, 12, 17
Waterhouse, John William *112*
waterspouts 21, 57–8, *58*, 116–18, *118*
Webster, John 8
Wesley, Charles 7
Wicker Man, The 18
Winge, Mårten Eskil *16*
witches 31
Wizard of Oz, The 149–50
Wragge, Clement 64
Wuthering Heights 119

Xerxes 69–70, *70*

Yoruba people 25

Zeus 11–12, 15